USING PHYSICAL SCIENCE

GADGETS & GIZMOS

GRADES 3–5

PHENOMENON-BASED LEARNING

Matthew Bobrowsky

Mikko Korhonen

Jukka Kohtamäki

USING PHYSICAL SCIENCE
GADGETS & GIZMOS
GRADES 3–5
PHENOMENON-BASED LEARNING

Arlington, Virginia

Claire Reinburg, Director
Wendy Rubin, Managing Editor
Andrew Cooke, Senior Editor
Amanda O'Brien, Associate Editor
Amy America, Book Acquisitions Coordinator

ART AND DESIGN
Will Thomas Jr., Director
Joe Butera, Senior Graphic Designer, cover and interior design

PRINTING AND PRODUCTION
Catherine Lorrain, Director

NATIONAL SCIENCE TEACHERS ASSOCIATION
David L. Evans, Executive Director
David Beacom, Publisher

1840 Wilson Blvd., Arlington, VA 22201
www.nsta.org/store
For customer service inquiries, please call 800-277-5300.

17 16 15 14 4 3 2 1

Cataloging-in-Publication Data
Bobrowsky, Matthew, 1955-
Using physical science gadgets and gizmos, grades 3-5 : phenomenon-based learning / by Matthew Bobrowsky, MSB Science, LLC, U.S.A., Mikko Korhonen and Jukka Kohtamaki, Ilmiopaja Oy, Finland.
1 online resource.
Includes bibliographical references and index.
Description based on print version record and CIP data provided by publisher; resource not viewed.
ISBN 978-1-936959-38-9 -- ISBN 978-1-938946-57-8 (e-book) -- ISBN 978-1-936959-38-9 1. Physics--Study and teaching (Primary)--Activity programs. 2. Physics--Study and teaching (Elementary)--Activity programs. 3. Physics--Experiments. I. Korhonen, Mikko, 1975- II. Kohtam?ki, Jukka, 1981- III. Title.
QC30
372.35--dc23

2014021986

C O N T E N T S

6

ELECTRIC CIRCUITS 51

Chapter 6 is organized a little differently from the other chapters because all the explorations involve different types of electrical circuits made from the same kit.

7

MAGNETISM 71

ABOUT THE AUTHORS

MATTHEW BOBROWSKY, PHD

Dr. Matt Bobrowsky has been involved in scientific research and science education for several decades. Currently at Delaware State University, he previously served as Director of the Physics Demonstration Facility at the University of Maryland—a collection of over 1,600 science demonstrations. Also at the University of Maryland, Matt was selected as a Faculty Mentor for the Fulbright Distinguished International Teachers Program, where he met Mikko Korhonen.

Matt has taught physics, astronomy, and astrobiology both in the classroom and online. He has written K–12 science curricula and serves on the Science Advisory Committee for the Howard County Public School System in Maryland. Matt has conducted countless professional development workshops for science teachers and special presentations for students, speaking on a variety of topics beyond physics, such as the scale of the universe, life in the universe, misconceptions about science among students and the public, the process of science, and science versus pseudoscience.

He is often asked to be an after-dinner speaker or keynote speaker at special events. Matt is a "Nifty Fifty" speaker for the USA Science & Engineering Festival and a Shapley Lecturer for the American Astronomical Society. Matt has received a number of awards for teaching excellence from the University of Maryland, including the Stanley J. Drazek Teaching Excellence Award (given to the top 2 instructors out of ~800) and the Board of Regents' Faculty Award for Excellence in Teaching (given to the top 3 instructors out of ~7,000). Matt's teaching is always innovative because he uses pedagogical techniques that are based on current science education research and known to be effective.

In his research, Matt has been involved in both theoretical and observational astronomy. He developed computer models of planetary nebulae—clouds of gas expanding outward from aging stars—and has observed them with telescopes on the ground as well as with the Hubble Space Telescope. One of the planetary nebulae that Matt investigated is the Stingray Nebula, which he discovered using Hubble.

MIKKO KORHONEN

Mikko Korhonen obtained a master's degree from Tampere Technical University in Finland, where he studied physics, mathematics, and pedagogics. Since then, he has been teaching physics, mathematics, and computer science at various schools in Finland. He has also developed a number of educational programs that brought some of his students to top scientific facilities in the world, including the Nordic Optical Telescope (NOT) observatory in La Palma, Spain, the CERN laboratory at the Franco-Swiss border, and the LATMOS laboratory in France. Most recently, some of his students have attended the Transatlantic Science School, which Mikko founded.

Mikko has written numerous other educational publications, including a book of physics experiments, manuals of physics problems with answers, an article on mathematics and logic for computer science, and two books with Jukka Kohtamäki on using toys to teach physics, one at the middle school and one at the high school level.

Mikko has obtained numerous grants and awards for his school and students, including awards from the NOT science school and the Viksu science competition prize, as well as individual grants from the Finnish National Board of Education and the Technology Industries of Finland Centennial Foundation, and grants for his "physics toys" project. His students are also award winners in the Finnish National Science Competition. Mikko received one of the Distinguished Fulbright Awards in Teaching, which brought him to the University of Maryland for a semester, where he worked with Matt Bobrowsky. Most recently, Mikko received the award of Distinguished Science Teacher in 2013 by the Technology Industries of Finland Centennial Foundation.

JUKKA KOHTAMÄKI

Jukka Kohtamäki obtained his master of science degree from Tampere University of Technology in Finland and since then has been teaching grades 5–9 at the Rantakyla Comprehensive School, one of the largest comprehensive schools in Finland. Jukka has participated in long-term professional development teaching projects and projects involving the use of technology in learning, as well as workshops that he and Mikko Korhonen conducted for Finnish science teachers. His writing includes teaching materials for physics and computer science, and he has written two books with Mikko on using toys to teach physics, one at the middle school and one at the high school level.

Jukka is a member of the group under the National Board of Education that is writing the next physics curriculum in Finland. He is also participating in writing curricula in chemistry and natural science (which is a combination of biology, geology, physics, chemistry, and health education). His goals are to get students engaged in lessons, to have them work hands on and minds-on, to encourage creativity in finding solutions, and to get students to discuss natural phenomena using the "language of physics." In 2013, Jukka received the Distinguished Science Teacher Award from the Technology Industries of Finland Centennial Foundation.

“The most beautiful thing we can experience is the mysterious. It is the source of all true art and science.”

— Albert Einstein

AN INTRODUCTION TO PHENOMENON-BASED LEARNING

The pedagogical approach in this book is called phenomenon-based learning (PBL), meaning learning is built on observations of real-world phenomena—in this case of some fun toys or gadgets. The method also uses peer instruction, which research has shown results in more learning than traditional lectures (Crouch and Mazur 2001). In the PBL approach, students work and explore in groups: Exercises are done in groups, and students' conclusions are also drawn in groups. The teacher guides and encourages the groups and, at the end, verifies the conclusions. With the PBL strategy, the concepts and the phenomena are approached from different angles, each adding a piece to the puzzle with the goal of developing a picture correctly portraying the real situation.

PBL is not so much a teaching method as it is a route to grasping the big picture. It contains some elements that you may have seen in inquiry-based, problem-based, or project-based learning, combined with hands-on activities. In traditional science teaching, it's common to divide phenomena into small, separate parts and discuss them as though there is no connection among them (McNeil 2013). In our PBL approach, we don't artificially create boundaries within phenomena. Rather, we try to look at physical phenomena very broadly.

PBL is different from project-based or problem-based learning. In project-based learning, the student is given a project that provides the context for learning. The problem with this is that the student is not necessarily working on the project out of curiosity but simply because they are required to by the teacher. To avoid having students view the project as a chore or just a problem that they have to solve, we employ PBL: The student's own curiosity becomes the driver for learning. The student explores not by trudging through a problem to get to the correct answer but by seeing an interesting phenomenon and wanting to understand what's going on. This works because interest and enthusiasm do not result from the content alone; they come from the students themselves as they discover more about a phenomenon. Personal experience with a phenomenon is always more interesting and memorable than a simple recitation of facts (Jones 2007; Lucas 1990; McDade 2013).

The goal in project-based learning is for the students to produce a product, presentation, or performance (Moursund 2013). PBL does not have that requirement; students simply enjoy exploring and discovering. This is the essence of science, and it is consistent with the philosophy of the *Next Generation Science Standards* (*NGSS*). Rather than simply memorizing facts that will soon be forgotten, students are doing real science. They are engaged in collaboration, communication, and critical thinking. Through this, students obtain a deeper understanding of scientific knowledge and see a real-world application of that knowledge—exactly what was envisioned with the *NGSS*. This is why, at the end of each chapter, we provide a list of relevant standards from the *NGSS*, further emphasizing our focus on the core ideas and practices of science, not just the facts of science.

The objective of PBL is to get the students' brains working with some phenomenon and have them discussing it in groups. A gizmo's functions, in most cases, also make it possible for teachers to find common misconceptions that students may harbor. It is important to directly address misconceptions because they can be very persistent (Clement 1982, 1993; Nissani 1997). Often the only way to remove misconceptions is to have students work with the problem, experiment, think, and discuss, so that they can eventually experience for themselves that their preconception is not consistent with what they observe in the real world.

> *"Most of the time my students didn't need me; they were just excited about a connection or discovery they made and wanted to show me."*
>
> —Jamie Cohen (2014)

We must also keep in mind that students can't build up all the scientific laws and concepts from scratch by themselves. Students will definitely need some support and instruction. When doing experiments and learning from them, the students must have some qualitative discussions (to build concepts) and some quantitative work (to learn the measuring process and make useful calculations). Experience with combining the two reveals the nature of science.

PBL encourages students to not just think about what they have learned but to also reflect on how they acquired that knowledge. What mental processes did they go through while exploring a phenomenon and figuring out what was happening? PBL very much lends itself to a **K-W-L** approach (what we **K**now, what we **W**ant to know, and what we have **L**earned). K-W-L can be enhanced by adding an H for "How we learned it" because once we understand that, we can apply those same learning techniques to other situations.

When you first look at this book, it might seem as if there is not very much textual material. That was intentional. The idea is to have more thinking by the students and less lecturing by the teacher. It is also important to note that the process of thinking and learning is not a race. To learn and really get the idea, students need to take time to think … and then think some more—so be sure to allow sufficient time for the cognitive processes to occur. For example, the very first experiment (using a toy car) can be viewed in just a few minutes, but in order for students to think about the phenomenon and really get the idea, they need to discuss the science with other group members, practice using the "language of science," and internalize the science involved—which might take 20 minutes. During this time, the students may also think of real-life situations in which the phenomenon plays a significant role, and these examples can be brought up later during discussions as an entire class.

HOW TO USE THIS BOOK

This book can be used in many ways. It can be used as a teacher's guide or as material for the students. In the hands of the teacher, the introductions and the questions can be used as the basis for discussions with the groups before they use the gadgets, that is, as a motivational tool. The teacher can ask where we see or observe the phenomenon in everyday life, what the students know about the matter prior to conducting the activities, and so on. The explorations can also be used to spark curiosity about a particular area of science and to encourage students to explore and learn.

Exactly how you present the material depends a lot on your students. Here's one approach: Have students work in groups. Studies have shown this to be a good way for them to learn. Have the students discuss with each other—and write down in their science notebooks—what they already know about the gadget and about the phenomenon it demonstrates. If they don't yet know what phenomenon the gadget shows, you can just have them carry out the steps provided for exploring that gadget. As they perform the exploration, they should write questions they think of about the gadget or the science involved. If the students are having trouble with this, you can get more specific and ask them, for instance, what they would need to know in order to understand what's going on; or ask where else they have seen something like this. Having students formulate questions themselves is part of PBL and also part of an inquiry approach. Asking questions is also how scientists start out an investigation. Be sure to give the groups plenty of time to attempt to answer their questions themselves.

It's great to let the students pursue the questions they raise and to encourage investigations into areas that they find interesting. This is part of "responsive teaching," and also part of PBL. If, after a good amount of time, the students are unable to come up with their own questions, you can start to present the questions in the book—but resist the temptation to just have students go down the list of questions. That is more structure than we want to have in view of our goal of presenting learning as exploration and inquiry. A good use of the questions would be to help guide your interaction with the groups as they explore a phenomenon.

Students can then work in groups to answer the questions, doing more experimentation as needed. The important point is that they won't learn much by simply being told an answer. Much more learning takes place if they can, through experimentation and reasoning, come up with some ideas themselves. Students may come up with an idea that is incorrect; rather than immediately correcting them, guide them toward an experiment or line of reasoning that reveals an inconsistency. It is not a bad thing if the students' first ideas are incorrect: This allows them to recognize that, through the process of science, it is possible to correct mistakes and come away with a better understanding—which is one of the main points of PBL.

While the student groups are investigating, you, the teacher, should be moving among them, monitoring conversations to determine whether the students are proceeding scientifically, for example, by asking questions or discussing ways to answer the questions—perhaps using the gadget, through debate, or even by doing web searches. This monitoring is part of the assessment process, as explained further in the Learning Goals and Assessment section.

After students have made a discovery or figured out something new, have them reflect on the mental process they went through to achieve that discovery or understanding. This reflection—sometimes referred to as metacognition—helps students recognize strategies that will be helpful for other challenges in the future. The combination of guidance and metacognition is consistent with a modern learning cycle leading to continuous increases in students' content knowledge and process skills.

LEARNING GOALS AND ASSESSMENT

The most important learning goal is for students to learn to think about problems and try a variety of approaches to solve them. Nowadays, most students just wait for the teacher to state the answer. The aim here is for students to enjoy figuring out what's going on and to be creative and innovative.

Combining this with other objectives, a list of learning goals might look something like this:

By the end of these lessons, students will

- think about problems from various angles and try different strategies;
- demonstrate process skills, working logically and consistently;
- collaborate with others to solve problems;
- use the language of science;
- reflect on the thinking processes that helped them to acquire new knowledge and skills in science; and
- view science as interesting and fun.

You will also notice that there are no formal quizzes or rubrics included. There are other ways to evaluate students during activities such as these. First, note that the emphasis is not on getting the "right" answer. Teachers should not simply provide the answer or an easy way out—that would not allow students to learn how science really works. When looking at student answers, consider the following: Are the students basing their conclusions on evidence? Are they sharing their ideas with others in their group? Even if a student has the wrong idea, if she or he has evidential reasons for that idea, then that student has the right approach. After all members of a group are in agreement and tell you, the teacher, what they think is happening, you can express doubt or question the group's explanation, making the students describe their evidence and perhaps having them discuss it further among themselves. Student participation as scientific investigators and their ability to give reasons for their explanations will be the key indicators that the students understand the process of science.

The PBL approach lends itself well to having students keep journals of their activities. Students should write about how they are conducting their experiment (which might differ from one group to another), ideas they have related to the phenomenon under investigation (including both correct and incorrect ideas), what experiments or observations showed the incorrect ideas to be wrong, answers to the questions supplied for each exploration, and what they learned as a result of the activity. The teacher can encourage students to form a mental model—perhaps expressed as a drawing—of how the phenomenon works and why. Then the students can update this model in the course of their investigations. Students' notebooks or journals will go a long way to helping the teacher see how the students' thinking and understanding have progressed. If there is a requirement for a written assessment, the journal provides the basis for that. As a further prompt for writing or discussion, encourage students to form a "bridge to the future" by asking questions such as "Where could we use this phenomena?" and "How could this be useful?" Students might also want to make a video of the experiment. This can be used for later reference as well as to show family and friends. Wouldn't it be great if we could get students talking about science outside the classroom?

A few of the questions asked of the students will be difficult to answer. Here again, students get a feel for what it's like to be a real scientist exploring uncharted territory. A student might suggest an incorrect explanation. Other students in the group might offer a correction, or if no one does, perhaps further experimentation, along with guidance from the teacher, will lead the students on the right course. Like scientists, the students can do a literature search (usually a web search now) to see what others know about the phenomenon. (Doing web searches also involves learning to recognize when a site is reputable and when it is not.) Thus there are many ways for a misconception to get dispelled in a way that will result in more long-term understanding than if the students are simply told the answer. Guidance from the teacher could include providing some ideas about what to observe when doing the experiment or giving some examples from other situations in which the same phenomenon takes place. Although many incorrect ideas will not last long in group discussions, the teacher should actively monitor the discussions, ensuring that students do not get too far off track and are on their way to achieving increased understanding. We've provided an analysis of the science behind each exploration to focus your instruction.

By exploring first and getting to a theoretical understanding later, students are working like real scientists. When scientists investigate a new phenomenon, they aren't presented with an explanation first—they have to figure it out. And that's what students do in PBL. Real scientists extensively collaborate with one another; and that's exactly what the students do here as well—work in groups. Not all terms and concepts are extensively explained; that's not the purpose of this book. Again, like real scientists the students can look up information as needed in, for example, a traditional science book. What we present here is the PBL approach, in which students explore first and are inspired to pursue creative approaches to answers—and have fun in the process!

PBL IN FINLAND

The Finnish educational system came into the spotlight after the Programme of International Student Assessment (PISA) showed that Finnish students were among the top in science literacy proficiency levels. In 2009, Finland ranked second in science and third in reading out of 74 countries. (The United States ranked 23rd and 17th, respectively.) In 2012, Finland ranked 5th in science and 6th in reading. (The United States ranked 28th and 24th, respectively.) Finland is now seen as a major international leader in education, and its performance has been especially notable for its significant consistency across schools. No other country has so little variation in outcomes among schools, and the gap within schools between the top- and bottom-achieving students is quite small as well.

Finnish schools seem to serve all students well, regardless of family background or socioeconomic status. Recently, U.S. educators and political leaders have been traveling to Finland to learn the secret of their success.

The PBL approach is one that includes responsive teaching, progressive inquiry, project-based learning, and in Finland at least, other methods at the teachers' discretion. The idea is to teach bigger concepts and useful thinking skills rather than asking students to memorize everything in a textbook.

AUTHORS' USE OF GADGETS AND GIZMOS

One of the authors (M.B.) has been using gizmos as the basis of teaching for many years. He also uses them for illustrative purposes in public presentations and school programs. The other two authors (M.K. and J.K.) have been using PBL—and the materials in this book—to teach in Finland. Their approach is to present scientific phenomena to students so that they can build ideas and an understanding of the topic by themselves, in small groups. Students progress from thinking to understanding to explaining. For each phenomenon there are several different viewpoints from which the student can develop a big-picture understanding as a result of step-by-step exploration. The teacher serves only as a guide who leads the student in the right direction. PBL is an approach that is not only effective for learning but is also much more fun and interesting for both the teacher and the students.

SAFETY NOTES

Doing science through hands-on, process, and inquiry-based activities or experiments helps to foster the learning and understanding of science. However, in order to make for a safer experience, certain safety procedures must be followed. Throughout this book, there are a series of safety notes that help make PBL a safer learning experience for students and teachers. In most cases, eye protection is required. Safety glasses or safety goggles noted must meet the ANSI Z87.1 safety standard. For additional safety information, check out NSTA's "Safety in the Science Classroom" at *www.nsta.org/pdfs/SafetyInTheScienceClassroom.pdf.* Additional information on safety can be found at the NSTA Safety Portal at *www.nsta.org/portals/safety.aspx.*

Disclaimer: Safety of each activity is based in part on use of the recommended materials and instructions. Selection of alternative materials for these activities may jeopardize the level of safety and therefore is at the user's own risk.

REFERENCES

Clement, J. 1982. Students' preconceptions in introductory mechanics. *American Journal of Physics* 50 (1): 66–71.

Clement, J. 1993. Using bridging analogies and anchoring intuitions to deal with students' preconceptions in physics. *Journal of Research in Science Teaching* 30 (10): 1241–1257.

Cohen, J. 2014. 18 ways to engage your students by teaching less and learning more with rap genius. *http://poetry.rapgenius.com/Mr-cohen-18-waysto-engage-your-students-by-teaching-less-andlearning-more-with-rap-genius-lyrics.*

Crouch, C. H., and E. Mazur. 2001. Peer instruction: Ten years of experience and results. *American Journal of Physics* 69 (9): 970–977.

Jones, L. 2007. The student-centered classroom. New York: Cambridge University Press. *www.cambridge.org/other_files/downloads/esl/booklets/Jones-Student-Centered.pdf.*

Lucas, A. F. 1990. Using psychological models to understand student motivation. In *The changing face of college teaching: New directions for teaching and learning*, no. 42, ed. M. D. Svinicki, 103–114. San Francisco: Jossey-Bass.

McDade, M. 2013. Children learn better when they figure things out for themselves: Brandywine professor's research published in journal. *PennState News. http://news.psu.edu/story/265620/2013/02/21/society-and-culture/children-learn-better-when-they-figure-things-out.*

McNeil, L. E. 2013. Transforming introductory physics teaching at UNC-CH. University of North Carolina at Chapel Hill. *http://user.physics.unc.edu/~mcneil/physicsmanifesto.html.*

Moursund, D. 2013. Problem-based learning and project-based learning. University of Oregon. *http://pages.uoregon.edu/moursund/Math/pbl.htm.*

Nissani, M. 1997. Can the persistence of misconceptions be generalized and explained? *Journal of Thought* 32: 69–76. *www.is.wayne.edu/mnissani/pagepub/theory.htm*.

ADDITIONAL RESOURCES

Bobrowsky, M. 2007. *The process of science: and its interaction with non-scientific ideas*. Washington, DC: American Astronomical Society. *http://aas.org/education/The_Process_of_Science.*

Champagne, A. B., R. F. Gunstone, and L. E. Klopfer. 1985. Effecting changes in cognitive structures among physics students. In *Cognitive structure and conceptual change*, ed. H. T. West and A. L. Pines, 163–187. Orlando, FL: Academic Press.

Chi, M. T. H., and R. D. Roscoe. 2002. The processes and challenges of conceptual change. In *Reconsidering conceptual change: Issues in theory and practice*, ed. M. Limón and L. Mason, 3–27. Boston: Kluwer Academic Publishers.

Dale, E. 1969. *Audio-visual methods in teaching*. New York: Holt, Rinehart, and Winston.

Donivan, M. 1993. A dynamic duo takes on science. *Science and Children* 31 (2): 29–32.

Enger, S. K., and R. E. Yager. 2001. *Assessing student understanding in science: A standards-based K–12 handbook*. Thousand Oaks, CA: Corwin Press.

Jacobs, H. H., ed. 2010. *Curriculum 21: Essential education for a changing world*. Alexandria, VA: ASCD.

McTighe, J., and G. Wiggins. 2013. *Essential questions: Opening doors to student understanding*. Alexandria, VA: ASCD.

Meadows, D. H. 2008. *Thinking in systems: A primer*. White River Junction, VT: Chelsea Green Publishing.

National Research Council (NRC). 2000a. *How people learn: Brain, mind, experience, and school.* Washington, DC: National Academies Press.

National Research Council (NRC). 2000b. *Inquiry and the National Science Education Standards: A guide for teaching and learning.* Washington, DC: National Academies Press.

National Research Council (NRC). 2012. *A framework for K–12 science education: Practices, crosscutting concepts, and core ideas.* Washington, DC: National Academies Press.

P-16 Science Education at the Akron Global Polymer Academy. Wait time. The University of Akron. *http://agpa.uakron.edu/p16/btp.php?id=wait-time.*

SPEED

On most roads there is a speed limit: You may have seen signs for them. On one road, the speed limit might be 45 miles per hour. Cars are not allowed to move faster than that. On wider roads, the speed limits are higher.

> **A few numbers:** *How fast can you run? The fastest runners in the world can run 100 meters in less than 10 seconds! This means they run at speeds of more than 10 meters per second, or 32 feet per second.*

The fastest pitchers in baseball can throw a ball at a speed of 100 miles per hour. How fast can *you* throw a ball?

So what do these numbers really mean? How do you know what your speed is? In the experiments in this chapter, you will learn all about speed.

1 Let's Explore!

SAFETY NOTE

Toy cars on the floor can be a serious trip or fall hazard. Use caution when working with equipment during this activity.

FIGURE 1.1: Constant Velocity Car

SPEED RACER

Here you will learn about constant speed, and you will measure some speeds with simple tools.

1. What toy do you have? How does it work?
2. Turn on the Constant Velocity Car (Figure 1.1) and then describe its speed. Is it faster or slower than a snail? a turtle? a rabbit? a horse? you?
3. *Speed* tells you how fast the toy or any object moves. Actually, it tells you how many meters the toy moves in one second—or how many miles a car travels in one hour. In your group, discuss what two things you need to measure if you want to find out the speed of the toy car.
4. Tell your teacher what you think about this. If your teacher believes you understand, then get the things that you need to measure the speed of the toy car.
5. Measure what you need to in order to find out the speed of the car. Repeat the measurements at least six times to make sure you did not make a mistake. Each measurement might be a little different. It is very important to write down your measurements. You can write down your measurements in a table like Table 1.1.
6. Compare the speed you calculated with the fastest speeds of some animals.

TABLE 1.1

TABLE FOR RECORDING MEASURED SPEEDS

Trial #	Distance in meters (m)	Time in seconds (s)	Divide the distance by the time. Write down your answer in meters/second (m/s).
1			
2			
3			
4			
5			
6			

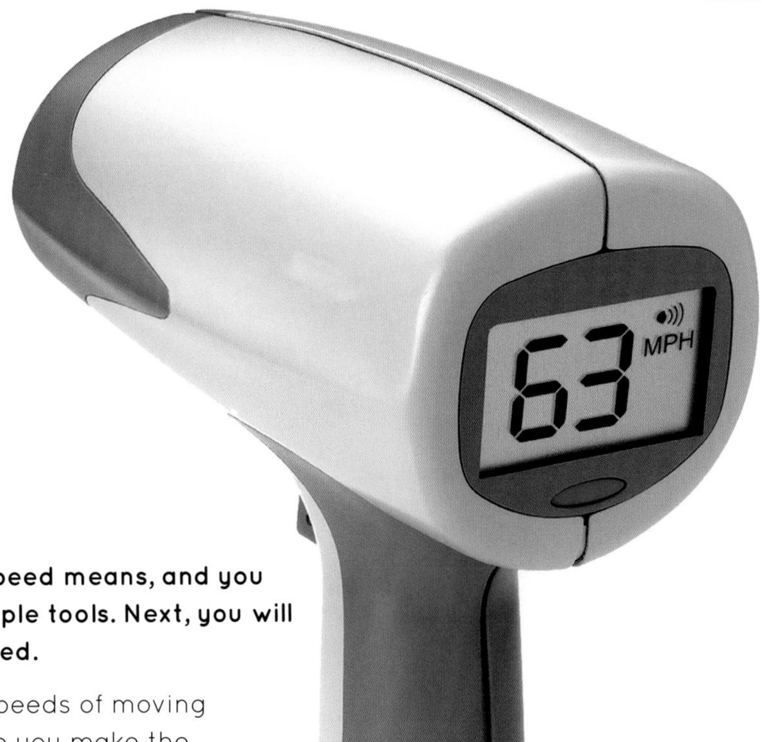

FIGURE 1.2: Velocity Radar Gun

MEASURING SPEED

Now you have an idea of what speed means, and you know how to measure it with simple tools. Next, you will see an easy way to measure speed.

1. Here you will estimate the speeds of moving objects. The gadget will help you make the measurement. What is the gadget? How does it work?
2. The Velocity Radar Gun (Figure 1.2) does not measure walking or running very well. (Arms and legs move in different directions at different speeds.) Discuss in your group what objects you might measure the speeds of with the radar gun in the school yard. Make a list.
3. For the objects in your list, write down how fast you think they move. Discuss this in your group and write down your guesses.
4. Then, make the measurements. Write down the speeds of the objects you chose to measure.
5. After you've measured all the speeds, look at your guesses. How close were they to the speeds that you measured?

FIGURE 1.3: Pull-Back Car

CHANGING SPEED

With the help of the Pull-Back Car (Figure 1.3), you will look at changes in speed.

1. Start by playing with the Pull-Back Car, which is a different kind of toy car from the Constant Velocity Car in the first experiment. How does this car work?
2. Describe how the car's speed changes as it moves.
3. Next, measure three speeds from different parts of the track. For this, you will need three stopwatches and a ruler for your group. Divide the car's track into three parts, each one meter long: 0–1 m, 1–2 m, and 2–3 m. Then, measure the time the car takes to travel each one-meter part of the track. Suggestion: Two students can work together to measure the time in each section. One student can use the stopwatch while the other one says "start" and "stop."
4. In which interval did the car move the fastest? the slowest? What can you say about how the car was moving?
5. Add a passenger or weights in the car. How does this change the motion of the car? Why is this?

What's Going On? 1

SPEED RACER

Speed tells you how fast an object is moving. To figure out the object's speed, you need to know two things: the distance traveled and the time it took. The distance can be measured in units such as inches, feet, meters, or kilometers. The time can be measured in units such as seconds or hours. You will need to use a stopwatch to measure how long the object takes to move. Once you know the distance and the time, you can figure out the speed. To get the speed, divide the distance by the time.

For example, if the distance is 10 meters and the time is two seconds you would divide:

10 meters ÷ 2 seconds = 5 meters/second

Your answer might not be a whole number. For example, if the distance is one meter and the time is 2.5 seconds, you would divide:

1 meter ÷ 2.5 seconds = 0.4 meters/second

FIGURE 1.4: Constant Velocity Car

Notice how when you divide meters by seconds, you get meters/second, which is pronounced, "meters per second." Meters/second is a unit of speed. Other units of speed are kilometers/hour or miles/hour. These are pronounced "kilometers per hour" and "miles per hour." You might have seen miles/hour written as MPH. On some roads, the speed limit is 55 MPH.

Table 1.2 shows speeds of different things and how long it takes them to move 40 yards on average.

The Constant Velocity Car (Figure 1.4) moves at a constant speed, which means it moves the same distance in each second.

TABLE 1.2

SPEEDS OF DIFFERENT THINGS AND THE TIME IT TAKES THEM TO MOVE 40 YARDS

Things timed	40 yard time in seconds (s)	m/s	MPH	km/h
Fast baseball pitch	0.8	44.7	100	161
Running back	4.5	8.0	18	29
Speed limit	1.3	29	65	105
Cougar	1.63	22.4	50	80.5
Rabbit	2.3	15.65	35	56.3
Turtle	73	0.5	1.1	1.8
Snail	2,727	0.013	0.03	0.048

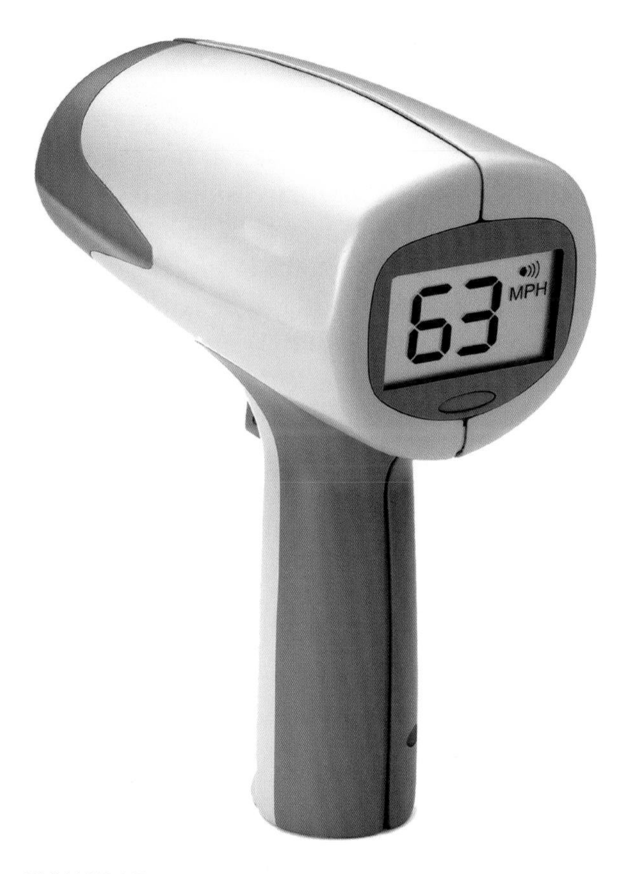

FIGURE 1.5: Velocity Radar Gun

MEASURING SPEED

In this experiment, we discussed speed. With the Velocity Radar Gun (Figure 1.5) you can measure speeds easily. This kind of gadget is used by the police to find out how fast cars are moving. Radar guns are also used at baseball games to measure the speeds of pitches.

CHANGING SPEED

In the last part of this chapter, you experimented with the Pull-Back Car (Figure 1.6). This is a car that changes speed. A car can change speed and it can change direction. A change in speed or a change in direction is called *acceleration*. Many people talk about acceleration only when something speeds up, but scientists will use the word acceleration for any change in speed or direction.

In your experiment, the car speeds up at first: It is accelerating. As the car accelerates, it takes less and less time to travel one meter. Then the car stops accelerating and moves at a constant speed for a short time. Finally, the car slows down.

You probably found out that adding a weight or a passenger to the car decreased the acceleration. The more weight an object has, the more difficult it is to speed up or slow down. When building a race car, the builders try to make the car as light as possible. That way it will speed up faster.

FIGURE 1.6: Pull-Back Car

Web Resources

Learn to graph position, velocity, and acceleration. Move a little man around with the mouse and plot his motion. Setting the position, velocity, or acceleration allows the simulation to move the man.

http://phet.colorado.edu/en/simulation/moving-man

Simulations to help students understand relationships among distance, speed, and time.

Information: *www.learnnc.org/lp/external/4413*

Simulation: *www.nctm.org/standards/content.aspx?id=25037*

Examples of the speed formula, a speed formula worksheet, and a study guide.

www.brighthubeducation.com/lesson-plans-grades-3-5/35416-teaching-the-speed-formula-includes-worksheet-and-study-guide/

A worksheet to help students determine the correct multiplication and division equations and calculate answers about distance and speed.

www.greatschools.org/worksheets-activities/5919-calculating-speed.gs

Relevant Standards

Note: The Next Generation Science Standards *can be viewed online at* www.nextgenscience.org/next-generation-science-standards.

PERFORMANCE EXPECTATIONS

K-PS2-2

Analyze data to determine if a design solution works as intended to change the speed or direction of an object with a push or a pull.

4-PS3-1

Use evidence to construct an explanation relating the speed of an object to the energy of that object.

DISCIPLINARY CORE IDEAS

PS2.A: Forces and Motion

- The motion of an object is determined by the sum of the forces acting on it; if the total force on the object is not zero, its motion will change. The greater the mass of the object, the greater the force needed to achieve the same change in motion. For any given object, a larger force causes a larger change in motion.
- All positions of objects and the directions of forces and motions must be described in an arbitrarily chosen reference frame and arbitrarily chosen units of size. In order to share information with other people, these choices must also be shared.

PS3.C: Relationship Between Energy and Forces

- A bigger push or pull makes things speed up or slow down more quickly. (secondary to K-PS2-1)

PS2.A: Forces and Motion

- Each force acts on one particular object and has both strength and a direction. An object at rest typically has multiple forces acting on it, but they add to give zero net force on the object. Forces that do not sum to zero can cause changes in the object's speed or direction of motion. (Boundary: Qualitative and conceptual, but not quantitative addition of forces are used at this level.) (3-PS2-1)

NASA
United States

2

FRICTION AND AIR RESISTANCE

Every day you experience *friction*. You notice it when walking, for example, or when slowing down while bicycling. Starting and stopping would be impossible without this important phenomenon.

Friction is very common. In some cases, there is only a little and in others there is a lot. Friction can be useful, or it can make things more difficult. When the floor has very little friction, we say it is slick. This idea is sometimes shown in comics with a banana peel that makes some unlucky person slip. Wet ice has very little friction, so it is slippery. Friction is useful when you don't want to slide. Many sports players wear sneakers: The rubber soles provide a lot of friction on most surfaces, preventing players from falling. Friction is also useful when you want to stop your bicycle. There is friction where the brakes rub on the wheels, and there is friction between the rubber tires and the road. All of this friction helps the bicycle stop when you apply the brakes.

You also experience *air resistance* when you ride your bicycle. You feel the air blowing against your face. This air hitting you and your bicycle might make it hard for you to go very fast. The air slowing you down is called *air resistance* or *drag*.

A push or a pull is called a *force*. Gravity is a force that pulls things down. Magnets cause a force that pulls on certain metals and pushes or pulls on other magnets. Friction is a force that can slow down a sliding object. Air resistance is also a force that can slow things down.

In the next experiments, you will explore these important everyday phenomena.

2 Let's Explore!

DRAGGING THE BLOCK

In these experiments, you will explore some things that affect the amount of friction. Have fun!

1. You have a Four-Sided Friction Block (Figure 2.1), a spring scale, and some kind of weight. The weight could be a book or anything else. What do you think you can find out with these things?
2. Tape down a sheet of paper.
3. First, choose one of the sides of the block and place it down on the paper.
4. Attach the spring scale to the block.
5. Put the weight on top of the block.
6. Practice making measurements with the spring scale and the block. The spring scale will tell you how strong a force is needed to move the block.
7. Start pulling the block with the scale. Do it gently and slowly! What is the highest reading on the scale before the block starts to move?
8. Now drag the block with the spring scale at a constant speed. Write down the reading on the scale.
9. Turn the block so that a different side is against the paper. Pull the scale again and write down the reading on the scale.

FIGURE 2.1: Four-Sided Friction Block

10. Repeat this with the other two sides of the block. Write down the reading on the scale each time.
11. Discuss with others in your group whether the readings are the same or different. Discuss why.
12. What is the role of friction in your life? Where do you find this phenomenon in everyday life? Where do you find it on your way to school?
13. Think of situations in which you would want a lot of friction. Think of situations in which it's useful to have less friction. Discuss your ideas with the rest of the class.

▼
SAFETY NOTE

Wear safety glasses or goggles.

FIGURE 2.2: Air Puck

SLIDING THE PUCK

With the Air Puck (Figure 2.2) you can study how objects act if there is very little friction.

1. Explore the gadget you were given. What is it? Find out how it works.
2. Take the Air Puck into the hallway. Make sure the Air Puck is turned off, then put it on the floor and give it a little push. See how far it goes.
3. Now turn the Air Puck on. Give it a little push again, just as hard as before. See how far it goes.
4. Was there a difference? Why or why not?
5. Discuss with others in your group where you experience instances of very little friction in real life.

SAFETY NOTE

- Wear safety glasses or goggles.
- The puck on the floor can be a serious trip or fall hazard. Use caution when working with equipment on the floor.

FIGURE 2.3: Running Parachute

▼
SAFETY NOTE

Watch for holes, trash, and other things that can cause slip, trip, or fall hazards while running on the ground.

IS RUNNING A DRAG?

Friction is a force that can slow something down. However, a skydiver cannot use friction to slow down. Skydivers must use something else—air resistance. Here you will use a Running Parachute (Figure 2.3), and you will not be jumping out of an airplane. In fact, you will not be jumping at all.

1. Put on the belt that is attached the parachute. Do a test run. If you can run into the wind, you will probably feel the air resistance even better. How does it feel to run with the parachute on? Why?
2. This parachute is horizontal, but the vertical kind that skydivers use works the same way. Can you explain how a parachute works?
3. Find a place where you can run at least 40 yards. Time your runs. First run that distance without the parachute and then run the 40 yards with it.
4. Discuss with others in your group whether there was a difference. Why or why not?

What's Going On? 2

DRAGGING THE BLOCK

With the Four-Sided Friction Block (Figure 2.4) you figured out some things about the phenomenon called friction. Friction pushes on things. That means it is a *force*. Remember that forces are pushes or pulls on objects.

FIGURE 2.4: Four-Sided Friction Block

Friction is a force that you notice when two objects' surfaces are in contact with each other. You might notice the friction more when one object is sliding against the other one. Do you see where there might be friction in the pictures below?

With the friction block you made some measurements to find out how the surface texture affects the friction. You can see the difference in the reading of the scale when you try your experiments. All four sides of the block are different, so the readings you got were different for each side. Perhaps you figured out that the surface texture is the key. The rougher the surface touching the paper, the more friction there is. For example, one side of the block has sandpaper on it, and sandpaper is very rough.

One more interesting thing is that the reading on the scale drops just as the block starts to move. This means that a greater force is needed to get the block moving than to keep it moving. Friction is stronger when the object is not moving.

Here is one more experiment: Start rubbing your palms together. Do it vigorously! If you do it hard

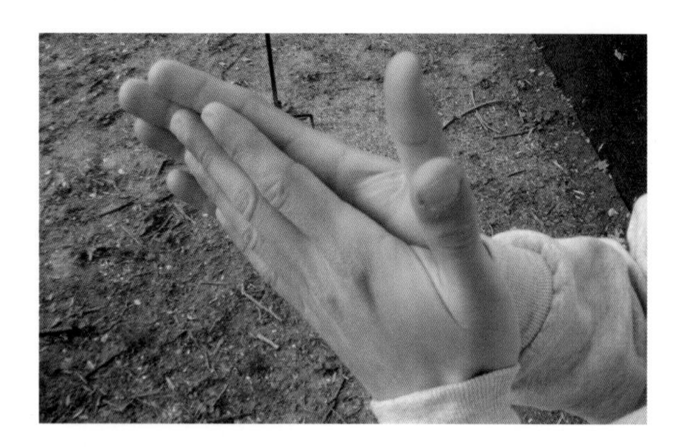

enough, you will notice something quite interesting—heat! Bring your heated palm close to your cheek. Can you feel the warmth? Heat is always produced if there is friction between sliding objects.

Sometimes you can also hear a sound caused by friction. For example, the friction from a violin bow causes a string to make a sound, and the tires of a car may squeal when it is speeding or breaking. Actually, they can squeal and smoke, showing that friction can cause both sound and heat.

Friction is often considered to be a problem. In many cases, people try to get rid of it or at least make it as small as possible. Oil can be used to reduce friction, which is why it is used in car engines or to fix a squeaky hinge.

The oil in the engine allows the parts of the engine to move more easily. Oil in a door hinge can get rid of the squeaky sound.

FIGURE 2.5: Air Puck

SLIDING THE PUCK

The Air Puck (Figure 2.5) slides on a cushion of air. Air has very little friction, so the puck slides easily. A puck also slides on air in an air hockey game.

FIGURE 2.6: Running Parachute

IS RUNNING A DRAG?

Air resistance can cause problems too. Air resistance is a force that occurs because moving objects hit the air in front of them and have to push the air out of the way. Air resistance must be taken into account when designing cars or airplanes. The car's design must be streamlined or sleek so most of the energy produced in the car's motor goes to the movement of the car—not pushing the air aside.

You may have noticed when you were running that the faster you run, the more the parachute (Figure 2.6) pulls back on you. You may also notice air resistance when running or riding your bike into the wind: In this case, you run into more air and therefore the air resistance is greater. After you timed your running with and without the parachute, you could calculate your speed. You can't run as fast with the parachute. However, air resistance can be useful when a skydiver uses a parachute.

Friction can also be very useful. Without the forces of friction, you would slip and slide every time you try to walk. You wouldn't be able to write with your pencil, light a match, or warm your hands by rubbing them together.

Web Resources

Design your own roller coaster, and decide how much friction there should be.
www.funderstanding.com/educators/coaster/

An experiment with friction: You're driving and there's a traffic jam up ahead. See if you can stop your vehicle right behind the traffic without running into it. Remember, when the road is wet, there's less friction and it will take longer to stop.
http://fearofphysics.com/Friction/friction.html

There is a patch of friction in the middle of the screen. Set the speed and amount of friction of a block and see what happens when it gets to the friction patch.
http://vnatsci.ltu.edu/s_schneider/physlets/main/fricpatch1.shtml

One mass sits atop another. The surface of one mass is frictionless, but there is friction between the two. There can be a push force applied to either mass. Set the masses, force, and amount of friction, and then "play" the animation.
http://vnatsci.ltu.edu/s_schneider/physlets/main/fric1.shtml

Perform Galileo's experiment with and without air resistance. Drop two different balls—or a ball and a feather—from the top of the Leaning Tower of Pisa. Drag the balls or a feather to Galileo's hands and then drop them.
Simulation: *www.planetseed.com/files/uploadedfiles/Science/Laboratory/Air_and_Space/Galileo_Drops_the_Ball/anim/en/index.html?width=740&height=570*
Teacher information: *www.planetseed.com/mathsolution/skydiving-air-resistance*

Relevant Standards

Note: The Next Generation Science Standards *can be viewed online at* www.nextgenscience.org/next-generation-science-standards.

The performance expectations in third grade help students formulate answers to questions such as: "How do equal and unequal forces on an object affect the object?"

PERFORMANCE EXPECTATION

3-PS2-1

Plan and conduct an investigation to provide evidence of the effects of balanced and unbalanced forces on the motion of an object.

DISCIPLINARY CORE IDEA

PS2.B: Types of Interactions

- Objects in contact exert forces on each other.

CROSSCUTTING CONCEPTS

Cause and Effect

- Cause and effect relationships are routinely identified. (3-PS2-1)
- Cause and effect relationships are routinely identified, tested, and used to explain change. (3-PS2-3)
- Cause and effect relationships are routinely identified and used to explain change. (5-PS1-4)

GRAVITY

You might have heard the phrase "What goes up, must come down." People say it a lot. Although some rockets do not come back down, most things you see do. You can see things go up and then come down every day.

Imagine you are holding a one-kilogram mass in your hand. Feel the force that is pulling the mass toward the ground. Then imagine holding a five-kilogram mass. You would probably notice that the force pulling down is stronger now. In fact, it is five times stronger. There is always a force between the Earth and other masses. This force is called *gravity* or *gravitation*.

In both of the imaginary experiments you just did, you felt the pull from a second mass—Earth's mass. That mass is so big that you can feel the gravitational force. The strength of the force depends on the masses of the objects.

How would it be if Earth had more mass? Less mass? What is gravity like on the Moon?

3 Let's Explore!

▼
SAFETY NOTE

- Wear safety glasses or goggles.
- Use caution when balls are on the floor. They can be a slip or fall hazard.

FALLING BALLS

How does gravity affect moving objects? Gravitation definitely affects objects on Earth. Is it possible to make a ball stay in the air longer? Here, you'll use the Vertical Acceleration Demonstrator (Figure 3.1) to find out.

1. Put the Vertical Acceleration Demonstrator at the edge of the table or on a stand so it is about 1.5 meters above the ground. Make sure that the gadget is balanced and level.
2. Put the balls in place, but do not yet release the spring.
3. Now decide: When the spring is released, which ball will hit the ground first? Or will the balls hit the ground at the same time?
4. Discuss this with other members of your group. See if your group can agree on what will happen—and why.
5. Pull the trigger of the Vertical Acceleration Demonstrator to find out what happens. *Hint*: Listen carefully!
6. The balls hit the ground. Is gravity still pulling on the balls when they lie on the floor?

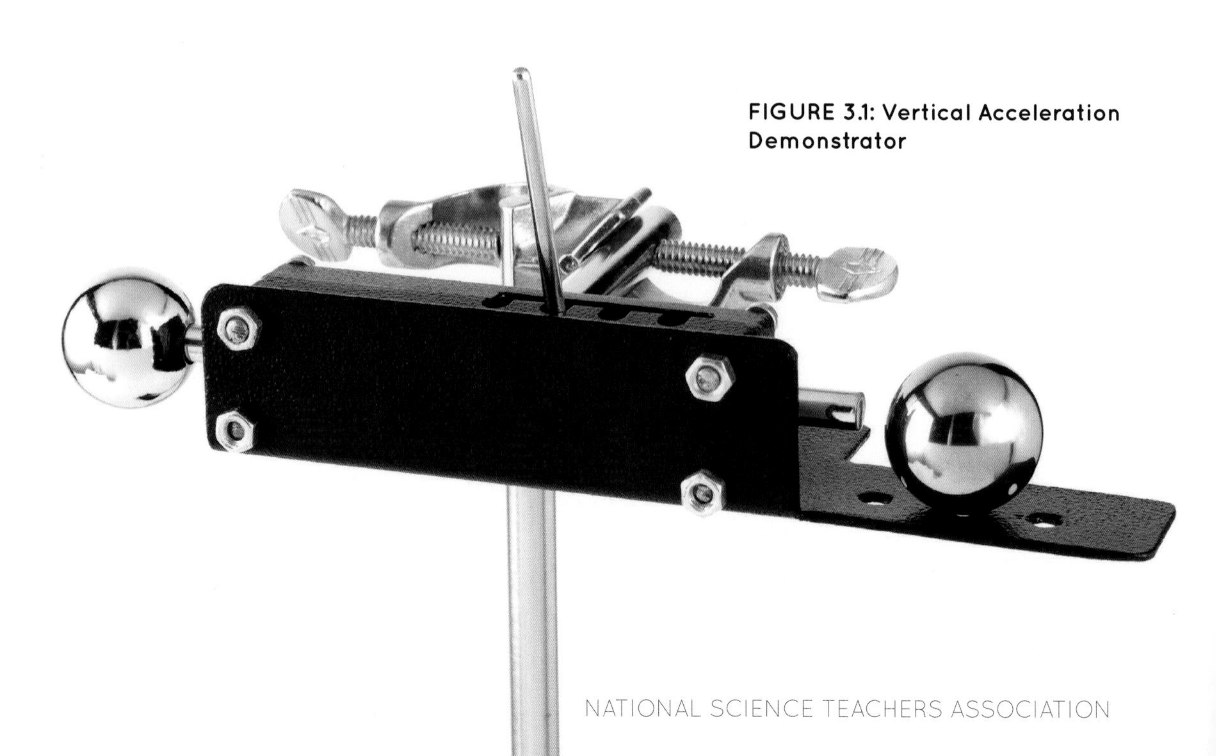

FIGURE 3.1: Vertical Acceleration Demonstrator

FIGURE 3.2: IR-Controlled UFO Flyer

SAFETY NOTES

- Wear safety glasses or goggles.
- Make sure all fragile items are removed from the area before launching the flyer.

PROPELLER PUZZLE

To take off, an object such as an airplane must fight gravity. To stay in the air, the force pushing upward must be as strong as the gravitational force. If the object stays at the same height in the air, the upward and downward forces are equal. To make an object that was not moving start to go higher, the upward force must be stronger than the gravitational force.

Gravitation tries to pull things toward the ground. There are many ways to fight against gravity. In the next experiment with the IR-Controlled UFO Flyer (Figure 3.2) you will explore one of them.

1. First, it is important to learn how to fly the IR-Controlled UFO Flyer.
2. Hold the flyer about two feet from the ground. Turn it on and release it.
    - Try to fly the flyer so that it hovers about as high as you are tall.
    - Now practice takeoffs and landings. When you can do both smoothly, you are ready to move to the next step.
3. Keep the flyer steadily in the air. Move your hand under the flyer as it hovers. What do you feel? Use this information to explain why the flyer stays in the air.
4. Next, cut out a circle of paper about the size of a CD. Put it on the table. Attach the flyer's landing skids to the paper.
5. Now, take off! What happens—and why?
6. Remove the paper.
7. Put a table in the middle of the classroom. First, set the flyer at the front of the classroom. Then, try to fly it over to land on the table.
8. Can you create some wind to make the flyer move sideways?
9. Try to lift a piece of adhesive or mounting putty with the flyer. Find out how much cargo it can lift. *Hint*: Attach the sticky tack to the bottom of the flyer.

▼
SAFETY NOTES

Wear safety glasses or goggles.

BALANCING BIRD

When an object is balanced, the forces in opposite directions are equal, and the object does not accelerate. In addition, there are at least two forces affecting it: One force is gravity pulling downward, and another is whatever force pushes or pulls upward and keeps the object from falling. You can balance an object by supporting it at only one point. It can be tricky, but in the next experiment you will learn how to do it.

1. First, try to balance one of your small schoolbooks on your fingertip. You need to find just the right spot. When you have found this spot, the book will be balanced.
2. Next, try to do the same for the Balancing Bird (Figure 3.3) you were given.
3. Why is the shape of the bird important for how you balance it?

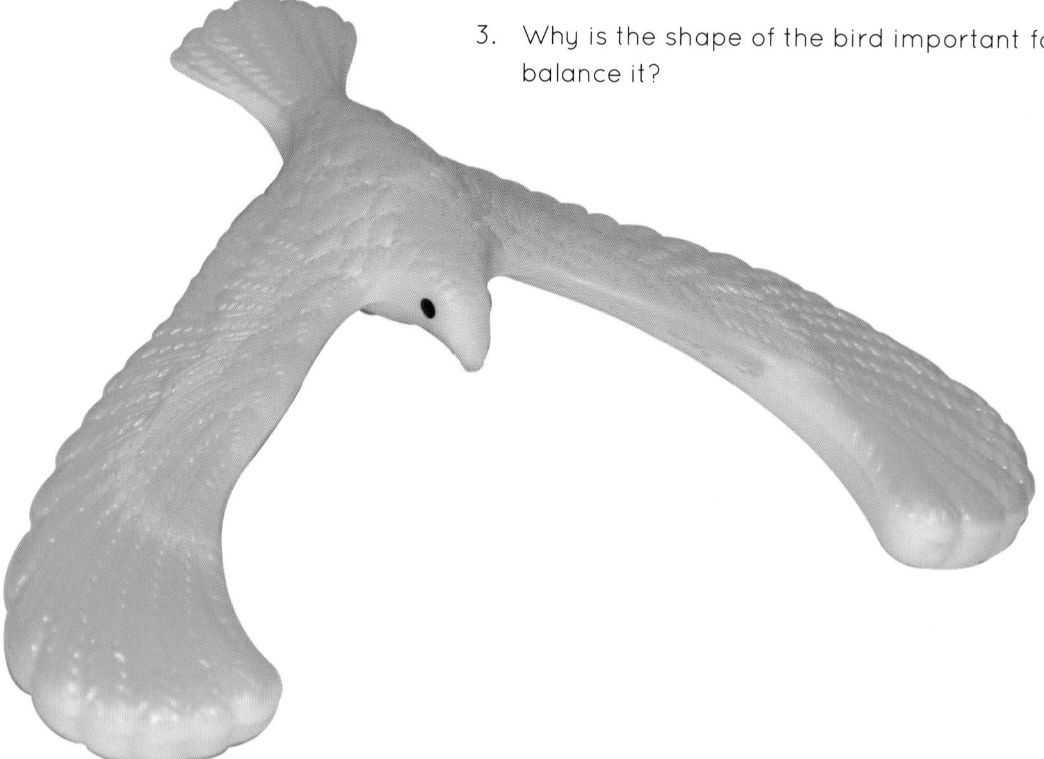

FIGURE 3.3: Balancing Bird

What's Going On? 3

FALLING BALLS

Gravitational force is caused by two masses affecting each other. You saw this in action with the Vertical Acceleration Demonstrator (Figure 3.4). Gravity between the Earth and the balls made the balls fall.

On TV or on the web you may have seen astronauts jumping on the Moon. The astronauts are not amazing jumpers, but with the Moon's weaker gravity, an astronaut can jump much higher there than on the Earth.

FIGURE 3.4: Vertical Acceleration Demonstrator

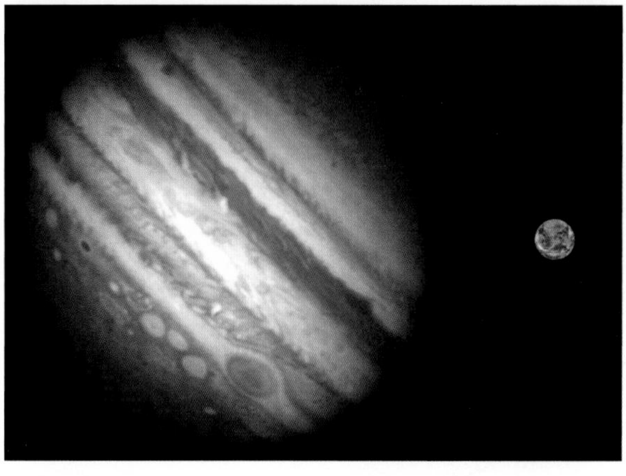

FIGURE 3.5: A photograph of Jupiter (on the left) is placed next to a photograph of Earth (on the right), so you can see how much larger Jupiter is.

These leaps are possible because the Moon has less mass than the Earth and therefore has weaker gravity than Earth.

Jupiter is the largest planet in the solar system (Figure 3.5). Its gravity is 2.6 times stronger than Earth's.

Gravity affects all objects on Earth. It doesn't matter if they are on the ground or in the air. There is also gravity under water. In the Falling Balls experiment, the downward (vertical) force from gravity is the same for both balls. The sideways (horizontal) movement does not change the time to fall, and the vertical motion does not affect the horizontal motion. A ball takes the same time to fall whether or not it is also moving sideways. In Figure 3.6, you can see how the balls stay side by side as they fall.

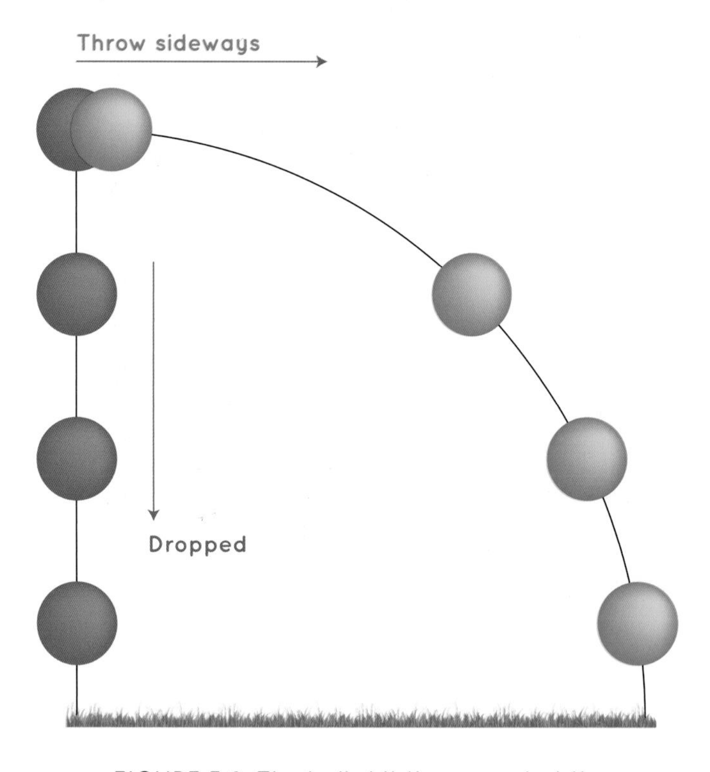

FIGURE 3.6: The balls hit the ground at the same time whether dropped vertically or thrown sideways.

PROPELLER PUZZLE

Imagine filling up a balloon with air and then letting it go. The air blows out one way, pushing the balloon in the opposite direction (Figure 3.7).

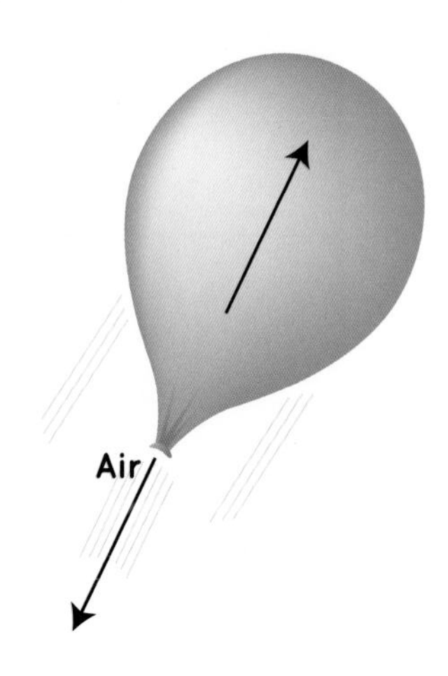

FIGURE 3.7: Releasing air out one end of the balloon pushes the body of the balloon in the opposite direction.

The IR-Controlled UFO Flyer (Figure 3.8) works in the same way. The flyer's propellers push air downward, and pushing the air downward causes a force upward—just like when you release a balloon. That force is as strong as the gravitational force downward. This is true when the flyer is hovering. When you attach paper to the flyer, the rotors' airflow pushes the paper *and* the flyer downward. This is why the flyer cannot take off with the paper attached, but a piece of sticky tack that is heavier than the paper can be lifted up with the flyer. Finally, although the flyer is at balance vertically, it can still move sideways with the wind.

FIGURE 3.8: IR-Controlled UFO Flyer

BALANCING BIRD

When an object is held up by its balance point, it stays in balance. You might imagine that the entire mass of the object is located at the balance point. It really is not, but it seems as if it is when you balance it. This balance point is sometimes called the *center of mass*.

The Balancing Bird's (Figure 3.9) balance point is on its beak. This is because weight was added at the end of the bird's wings, which are spread wide and extend forward. The weights in the tips of the bird's wings spread out the bird's weight so that its center of mass is at its beak.

FIGURE 3.9: Balancing Bird

You can also show a balance point with an empty shoe box, weights or stones, and a table. Put some weights in one side of the box. Then see if you can slide the box over so that most of it is beyond the edge of the table. The box is still in balance because the balance point is still on the table (Figure 3.10).

Another thing that affects the balance of an object is the amount of supporting area. For example, the larger the area between the four legs of a chair, the better the chair stays in balance and doesn't tip over. In fact, the Balancing Bird is also balanced when lying on the table—the supporting area is then bigger than just the beak. Can you find the balancing points in the picture below?

FIGURE 3.10: The cardboard box with weights balances as long as the balance point is on the table.

Web Resources

Animation of balls being dropped and thrown sideways.
www2.honolulu.hawaii.edu/instruct/natsci/science/brill/sci122/Programs/p14/projectile.mov

A classroom activity that simulates landing on a moving target.
www.pbs.org/wgbh/nova/education/activities/2110_carrier.html

A classroom activity to investigate center of mass (or center of gravity).
www.pbs.org/wgbh/nova/education/activities/2403_sle1ston.html

A classroom activity to investigate cause and effect and center of gravity (mass) of objects. (Click on "Download" button.)
http://on.docdat.com/docs/2733/index-21080.html

Relevant Standards

Note: The Next Generation Science Standards *can be viewed online at* www.nextgenscience.org/next-generation-science-standards.

PERFORMANCE EXPECTATIONS

3-PS2-1

Plan and conduct an investigation to provide evidence of the effects of balanced and unbalanced forces on the motion of an object.

5-PS2-1

Support an argument that the gravitational force exerted by Earth on objects is directed down. [Clarification Statement: "Down" is a local description of the direction that points toward the center of the spherical Earth.] [*Assessment Boundary: Assessment does not include mathematical representation of gravitational force.*]

DISCIPLINARY CORE IDEAS

PS2.A: Forces and Motion

Each force acts on one particular object and has both strength and a direction. An object at rest typically has multiple forces acting on it, but they add to give zero net force on the object. Forces that do not sum to zero can cause changes in the object's speed or direction of motion. (Boundary: Qualitative and conceptual, but not quantitative addition of forces are used at this level.)

PS2.B: Types of Interactions

Objects in contact exert forces on each other.

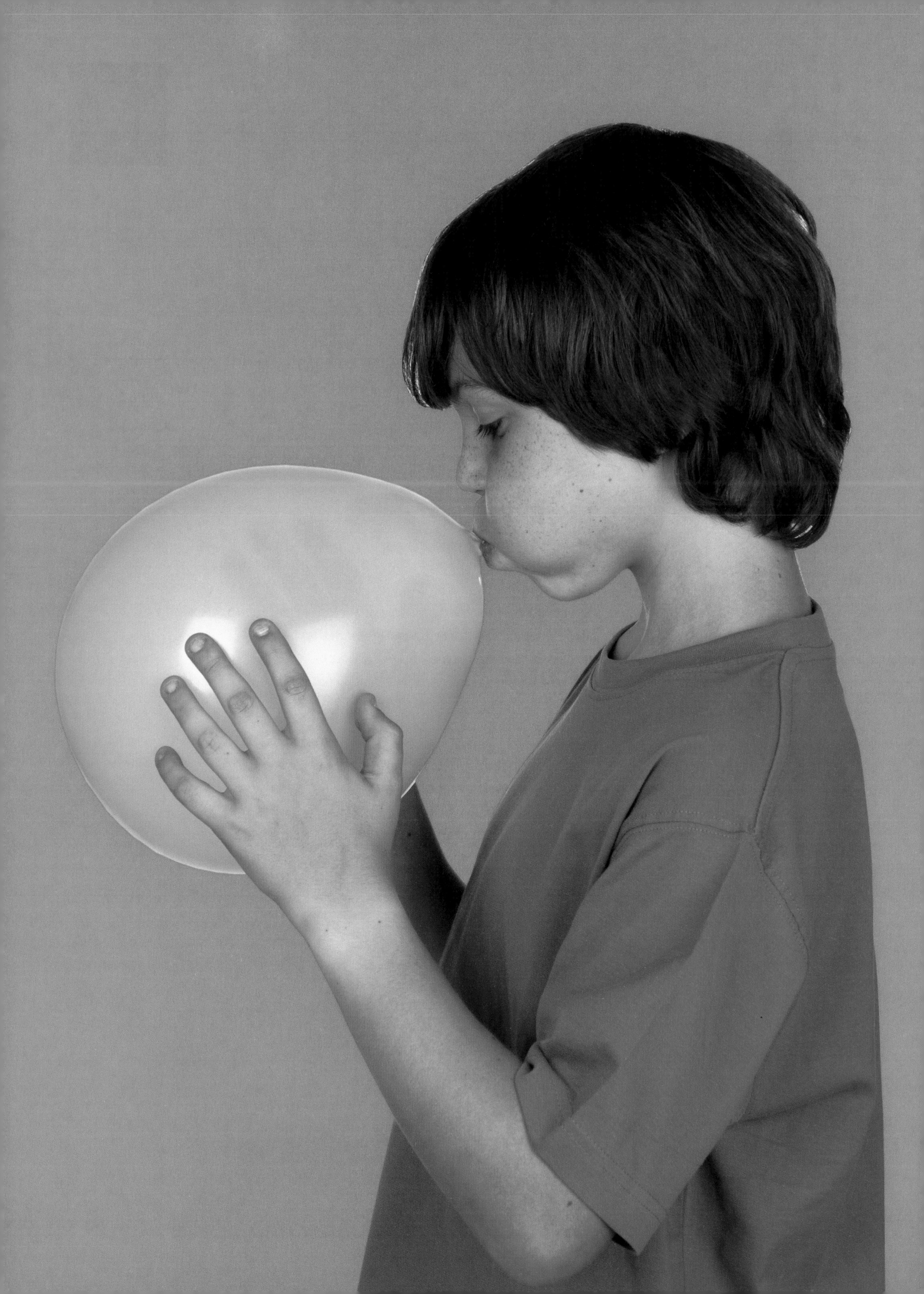

AIR PRESSURE

As you learned earlier, gravity affects all objects. This includes air molecules—the small particles that make up air. Gravity pulls the molecules toward the ground, and the molecules push on the ground. We call that pushing *air pressure* or *atmospheric pressure.* Near the ground there are more molecules and they are squeezed together more tightly, the air pressure is higher near sea level and lower up on a high mountain.

Air pushes in all directions because air molecules are constantly moving in all directions. The molecules bump into, or collide, with each other and with everything they touch. In a container, the air molecules collide with the walls. These collisions cause a force that pushes the walls outwards: You see this when the rubber expands as you blow up a balloon.

In the following experiments you will learn about air pressure. You may have seen terms such as *high pressure, excess pressure, low pressure,* and *underpressure*. Let's find out what they mean!

4 Let's Explore!

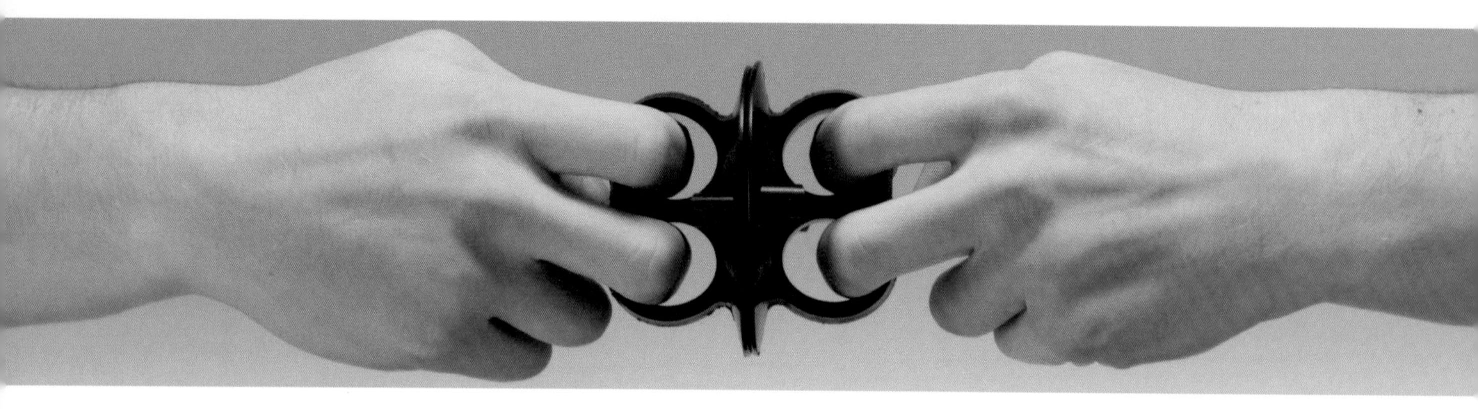

FIGURE 4.1: Atmospheric Pressure Cups

▼

SAFETY NOTES

- Wear safety glasses or goggles.
- To prevent injury, make sure arms are in an area free of any objects when pulling the suction cups apart.

IMPRESSIVE PRESSURE

We do not really feel the atmospheric pressure and the forces it creates. One reason for this is that we are used to these forces. Another reason is that the pressure is pushing equally on all parts of us, inside and out. In the following experiments with the Atmospheric Pressure Cups (Figure 4.1), you will get an idea of how strong the forces are.

1. What are the gadgets you have? How do they work?
2. Compress the cups against each other. Then pull the cups apart. They are probably very tight, so be careful!
3. There is an easier way to separate the two halves. Can you discover it?
4. Place a piece of cloth between the cups.
5. How do the cups work if there is some cloth between them? Why is that?
6. Try lifting different objects using the pressure cups. Take notes on things you can and cannot lift with them.
7. How does the kind of surface affect how well the pressure cups work? What is the same about the materials you are able to lift with the pressure cup? What is the same about the materials you are not able to lift?
8. Why do the two pressure cups stay together?

PRESSURE POWER

Air pressure squeezes every object. Let's see what happens if we remove the pressure with the Vacuum Pumper and Chamber (Figure 4.2).

1. You have a container and a pumper. Find out how the pump works.
2. Before attaching the pump to the container, find out which way the air moves while you are pumping. Which way is it?
3. You were given something to put in the container. It might be a balloon, a half-filled water balloon, a marshmallow, shaving foam, whipped cream, or something else.
4. What do you see in each of these cases? What happens while you pump? Why?

SAFETY NOTE

- Wear safety glasses or goggles.
- Never stand over or near the water rocket while it is being pressurized or launched.
- Stay clear of the stake—it presents a potential impalement hazard.
- Only perform this experiment outside in an open field, never inside.

FIGURE 4.2: Vacuum Pumper and Chamber

SAFETY NOTES

- Wear safety glasses or goggles.
- Before launching the rocket, make sure everyone is at a safe distance.
- Never stand over or near the water rocket while it is being pressurized or launched.
- Stay clear of the stake—it presents a potential impalement hazard.
- Only perform this experiment outside in an open field, never inside.

ROCKET!

This experiment must be done outdoors in a field with open sky. You will see the effect of gravitation and excess pressure. It may be surprising how high the Air-Powered Projectile (Figure 4.3) rises, so before launching the rocket, the area nearby must be made safe.

1. First, your teacher will launch the Air-Powered Projectile straight up. After the rocket lands, look at it closely. Explain what happened.
2. Then, select one of the wedges (triangular blocks) and place it under the launch pad. Try to guess where the rocket will land. Each group can estimate its own landing spot and mark it on the ground before the rocket is launched again.
3. Which angle of the launch pad do you think will make the rocket fly the farthest distance? Test your predictions.

FIGURE 4.3: Air-Powered Projectile

What's Going On?

TABLE 4.1

PRESSURE IN DIFFERENT PLACES IN DIFFERENT UNITS

When measuring pressure, several different units can be used: atmospheres (atm), bars (bar), kiloPascals (kPa), and pounds per square inch (PSI).

	Units			
Pressure location	**atm**	**bar**	**kPa**	**PSI**
Air pressure at sea level	1	1.01	101	14.7
Car tire	3.18	3.22	322	46.7
Bike tire	5.08	5.15	515	74.7
Basketball	1.54	1.57	157	22.7
Air at top of Mount Everest, 8,800 meters above sea level	0.332	0.337	34	5
Mars atmosphere	0.00592	0.00600	0.600	0.087
Venus atmosphere	90.5	91.7	9170	1330

Additional info: For the car tire, bike tire, and basketball, the reading on the pressure gauge is the amount by which the pressure is greater than the surrounding air pressure of 14.7 PSI. For example, the pressure gauge on the bike tire will read 60 PSI, while the actual pressure in the tire is 74.7 PSI.

IMPRESSIVE PRESSURE

With the Pressure Cups (Figure 4.4), it's easy to explore the forces caused by atmospheric pressure. Pressing the cups together squeezes the air out from between them. When you start to pull the cups apart, the space between the cups gets larger, but the number of air molecules remains the same, meaning the molecules are now more spread out. As a result, the molecules don't hit the cups as often, so the pressure between the cups is lower. The air pressure outside the cups is still the same—higher than the pressure inside—so it pushes the cups together.

Letting the air leak out from between the cups equalizes the pressure between the inside and the outside of the cups. With cloth between the cups, air molecules leak out because they can pass through the cloth. This allows the pressure inside and outside of the cups to be the same. This is also the case when the surface of an object you are trying to lift is rough. All the things that the cups can lift are smooth. All the things that the cups won't lift are rough.

FIGURE 4.4: Pressure Cups

As an extension, note that the same ideas about pressure apply to any fluid (liquid or gas), not just air. For instance, think about how an octopus attaches itself to glass under water.

PRESSURE POWER

With the Vacuum Pumper and Chamber (Figure 4.5) it is easy to observe the effects that air pressure has. To see what air pressure can do, you need to create differences in pressure. That means making one place have less air pressure than another place. With the chamber and pump, it is easy and safe to do.

You probably noticed that the pump makes air come out of the container. It is working in the reverse manner of a normal pump. Instead of adding air, as a pump does when used to add air to a bicycle tire, this pump makes the air move out of the chamber.

Molecules colliding in the chamber cause the air pressure there. The air pressure in the chamber is lower because you removed some air from it. With fewer molecules, there are also fewer collisions. Fewer collisions means lower pressure. This lower pressure is sometimes called *underpressure*, meaning that the pressure inside is lower than the normal, outside air pressure.

You also noticed that items that contain air, such as marshmallows, air balloons, or different kind of foams tend to expand in the chamber. Let's discuss the marshmallow demonstration first. When you start pumping, you remove the air from around the marshmallow. The air pressure is lower now: It doesn't push as hard on the marshmallow. When the pressure in the chamber is lower, the small air bubbles trapped inside the marshmallow start to expand because the air pressure from those air bubbles is greater than the air pressure outside the marshmallow. When you let the air flow back to the chamber, the marshmallow shrinks to a smaller size than when it started. That's because when it expanded, some of its stretchy parts got broken, causing it to collapse to a smaller size.

FIGURE 4.5: Vacuum Pumper and Chamber

With the other foams or a balloon, something similar happens. The air trapped inside the foam or balloon expands when the pressure in the chamber is low.

Question for discussion: What makes dirt go into a vacuum cleaner?

ROCKET!

The two experiments involved underpressure. With the rocket, you saw what can happen if there is excess pressure.

The pressure in the rocket's launch tube gets higher when you pump air into it. The black plastic cover will hold a certain amount of pressure. Once the pressure exceeds that amount, the cap pops and the compressed air comes out and expands, launching the rocket. The compressed air pushes the rocket up.

The Air-Powered Projectile (Figure 4.6) will fly pretty far. With different plastic covers, you can adjust the launching pressure and the distance that the rocket flies. The rocket flies farthest when the pressure is highest and when the launch angle is about 45 degrees.

Air compressed in a balloon pushes on the balloon when you let it go. Do you see how the balloon is like a rocket?

Large rockets do not have air coming out the back, but use other gases. Rockets will work with any kind of compressed gas. The photo to the right shows the rocket that first brought men to the moon in 1969. Would you like to ride in a rocket?

FIGURE 4.6: Air-Powered Projectile

Web Resources

An activity that shows students air pressure at work.
www.canteach.ca/elementary/physical6.html

Just as in the rocket exploration, students learn about projectile motion by firing various objects in this simulation. You can set the angle, initial speed, and mass, and add air resistance. Make it a game by trying to hit a target.
Info: *http://phet.colorado.edu/en/simulation/projectile-motion*
Simulation: *http://phet.colorado.edu/sims/projectile-motion/projectile-motion_en.html*

Marshmallow man in a vacuum.
www.videojug.com/film/marshmallow-man-in-a-vacuum

Air pressure pushes an egg into a bottle.
www.youtube.com/watch?v=crWA2FkmHnI

A slow-motion version of the egg being pushed into the bottle.
www.youtube.com/watch?v=OqKwxytsvvQ&noredirect=1

Shaving cream in vacuum
physics.wfu.edu/demolabs/demos/avimov/fluids/shaving_cream_vacuum/shaving_cream.MPG

Relevant Standards

Note: The Next Generation Science Standards *can be viewed online at* www.nextgenscience.org/next-generation-science-standards.

The performance expectations in third grade help students formulate answers to questions such as: "How do equal and unequal forces on an object affect the object?"

[In the performance expectations for all elementary grades], students are expected to demonstrate grade-appropriate proficiency in asking questions and defining problems; developing and using models, planning and carrying out investigations, analyzing and interpreting data, constructing explanations and designing solutions, engaging in argument from evidence, and obtaining, evaluating, and communicating information.

PERFORMANCE EXPECTATION

3-PS2-1

Plan and conduct an investigation to provide evidence of the effects of balanced and unbalanced forces on the motion of an object.

DISCIPLINARY CORE IDEA

PS2.A: Forces and Motion

- Each force acts on one particular object and has both strength and a direction. An object at rest typically has multiple forces acting on it, but they add to give zero net force on the object. Forces that do not sum to zero can cause changes in the object's speed or direction of motion. (Boundary: Qualitative and conceptual, but not quantitative addition of forces are used at this level.)

CROSSCUTTING CONCEPT

[Also in all elementary grades:] The crosscutting concepts of patterns; cause and effect; energy and matter; systems and system models; interdependence of science, engineering, and technology; and influence of engineering, technology, and science on society and the natural world are called out as organizing concepts for these disciplinary core ideas.

Cause and Effect

- Cause and effect relationships are routinely identified.

ELECTRICITY

Have you ever rubbed a balloon on your clothes or your hair? What happened? Perhaps you noticed that the balloon somehow made your hair stand up. Maybe you could then stick the balloon to a wall or a ceiling. This phenomenon is called *static electricity*.

The same phenomenon occurs when you touch a doorknob and get a shock. This happens because your shoes and clothes gather electric charges. When the difference in charges becomes high enough, you can make a small spark. Clouds can also gather electric charges. They make a big spark that we call lightning.

Now you will do some experiments so you can learn more about static electricity and electric charges.

5 Let's Explore!

SAFETY NOTE

Wear safety glasses or goggles.

FIGURE 5.1: Fun Fly Stick

FUN FLY STICK

With the Fun Fly Stick (Figure 5.1), you can make science look like magic. The phenomenon demonstrated here is something you may have experienced before.

1. Learn how the Fun Fly Stick works. Do you see how to turn it on and off?
2. Find the butterfly flyer and put it on the end of the Fun Fly Stick.
3. Press the button on the stick. What happens?
4. Shake the butterfly loose from the Fun Fly Stick. Can you make the butterfly fly around the classroom? What happens if the flying butterfly touches a wall or another student? (Note that you do not have to keep the button pressed while you are making the butterfly fly around.)
5. What keeps the butterfly in the air? Touch the butterfly gently with your finger. Does this give you an idea?
6. Select some other flyers and make them fly. Discuss with others why the flyers do what they do.
7. Next, you can try to make electric wallpaper. Depending on the weather and the kind of wall in your classroom, this might not work—but you can still try. Hold a sheet of paper against a wall and slide the Fun Fly Stick along the paper. Do this as though you were ironing the paper against the wall. Try to make the paper stick to the wall. If it sticks to the wall, talk about why it did.

LIGHTNING GLOBE

In the last experiment, you saw how electric charges sometimes repel and sometimes attract each other. There are two kinds of electric charges: positive and negative. When charges are the same kind, such as two positive charges, they push each other apart. When two charges are different kinds, one positive and one negative, they attract each other. In this experiment, you will use the Plasma Globe (Figure 5.2) to explore some more about electric charges and the phenomenon they cause called static electricity.

SAFETY NOTES

- Wear safety glasses or goggles.
- As with all electrical equipment, make sure the lightning globe is clear of any water sources.

1. Plug in the globe and turn it on. The gadget has a switch for three different options. Choose the one that keeps the globe on continuously.
2. Explore the Plasma Globe. Touch the top with just one finger.
3. Next, touch the globe with two fingers and then with the palm of your hand. What do you see? Discuss with others what happened and why.
4. Keep one finger on one side of the globe. How does the spark move? Discuss why. How is what is happening in this experiment similar to lightning?
5. Put your finger very close to the surface of the globe. Can you make a small spark between the globe and your finger? If you did that, you might have been able to smell a gas called *ozone*. Sparks of static electricity create ozone.
6. Now let your teacher take a fluorescent light tube and bring it close to the Plasma Globe. Discuss what happened with others in your class.

FIGURE 5.2: Plasma Globe

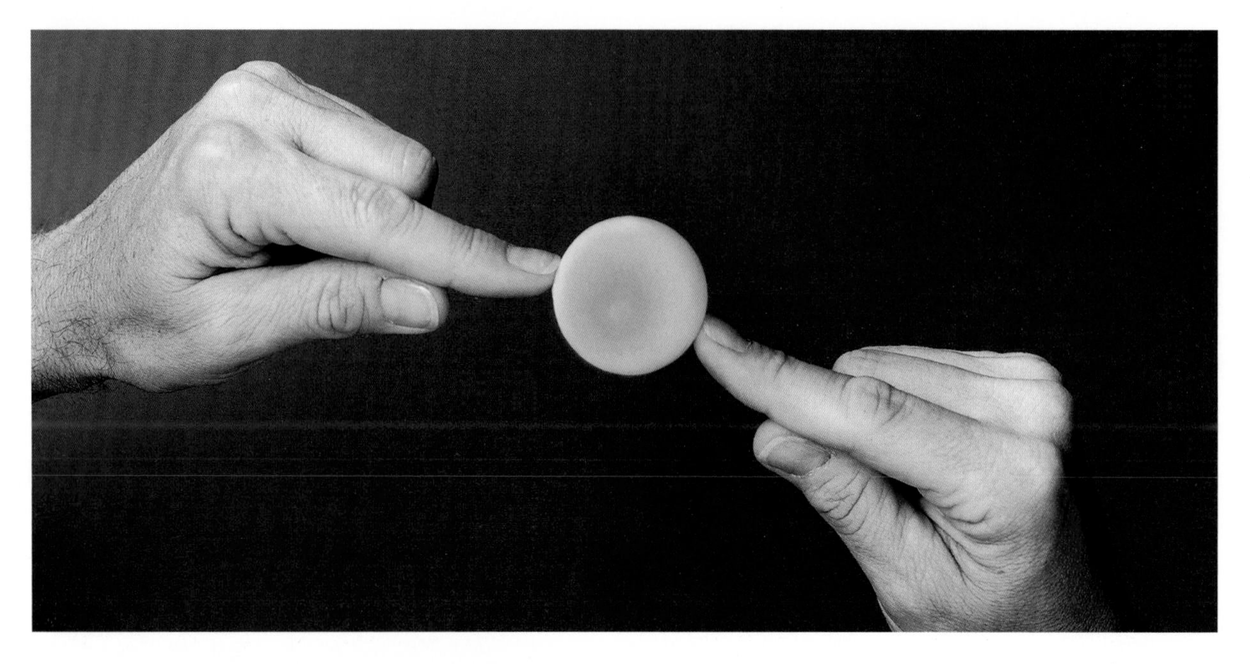

FIGURE 5.3: Energy Ball

HAVING A BALL

You have become acquainted with electric charges. In this experiment, the Energy Ball (Figure 5.3) will help you make them go around a *circuit*. A circuit can be a loop or a circle.

1. Find out what the Energy Ball does.
2. Make a circle of students in the classroom. Everyone in the circle must hold hands or touch elbows. Give the Energy Ball to two students in the circle so that each of them touches one of the metal strips.

▼
SAFETY NOTE

Wear safety glasses or goggles.

3. You'll know that your group has made a *closed circuit* when the Energy Ball makes a sound.
4. What can you do to make the Energy Ball stop making a sound? That is called *opening the circuit*.
5. Now close the circuit again. Did the Energy Ball start making a sound again?
6. Discuss with the class the idea of open and closed circuits.
7. How many students can form a closed circuit so that the Energy Ball still works?

What's Going On? 5

FUN FLY STICK

All matter consists of atoms. Atoms are very tiny building blocks for all materials. Everything you see around you is made of atoms. In an atom there is a part at the center called the *nucleus*. Surrounding the nucleus are *electrons* (Figure 5.4). The nucleus has a positive charge. The electrons have negative charges.

If an object has a negative charge, it has been given more electrons. If an object has a positive charge, the object has lost electrons. Electrons can move from one object to another if you rub two objects against each other. When you rub a balloon against your hair, both your hair and the balloon become charged—oppositely charged.

The objects that have opposite charges attract each other. You see this when you bring the charged balloon close to your hair (Figure 5.5). Your hair gets attracted to the balloon and sticks out toward it.

Your charged hair might stick out in other directions too. The charges on your hair are trying to get as far from each other as possible. There might even be enough charges on the balloon and your hair to make the balloon stick to your hair without you holding it (Figure 5.6).

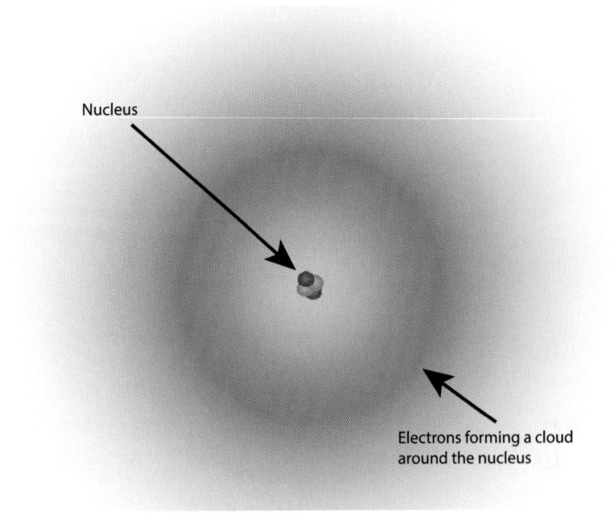

FIGURE 5.4: A helium atom with its nucleus and electrons. The electrons form a cloud around the nucleus.

When two things have the same charge, they push each other apart. We say that they *repel* each other. With the Fun Fly Stick (Figure 5.7, p. 46), both the flyers and the stick have positive charges.

FIGURE 5.5: When you bring a charged balloon near your head, it causes your hair to stick out.

FIGURE 5.6: With enough charge, the balloon will stick to your head without you holding it.

FIGURE 5.7: Fun Fly Stick

They got the same charge when they touched each other. Since they have the same kind of charge, they repel each other, and the flyer gets pushed away from the stick. This is why the flyer stays in the air above the stick.

The flyer gets the same positive charge all over it, making it spread out as the different parts try to get as far away as they can from each other.

LIGHTNING GLOBE

The Plasma Globe (Figure 5.8) is a very interesting gadget. It works because of a big difference in charge between two parts of the globe. A big difference in charge is called a *voltage*. (Where there are dangerous amounts of electricity, you might see a sign that says, "Danger. High voltage.")

In the middle of the globe there are many charges that repel each other. They try to move farther from the center to get away from the each other. Putting a finger on the globe creates an easier path for the charges to follow as they move farther away. This is why you see a bigger spark.

FIGURE 5.8: Plasma Globe

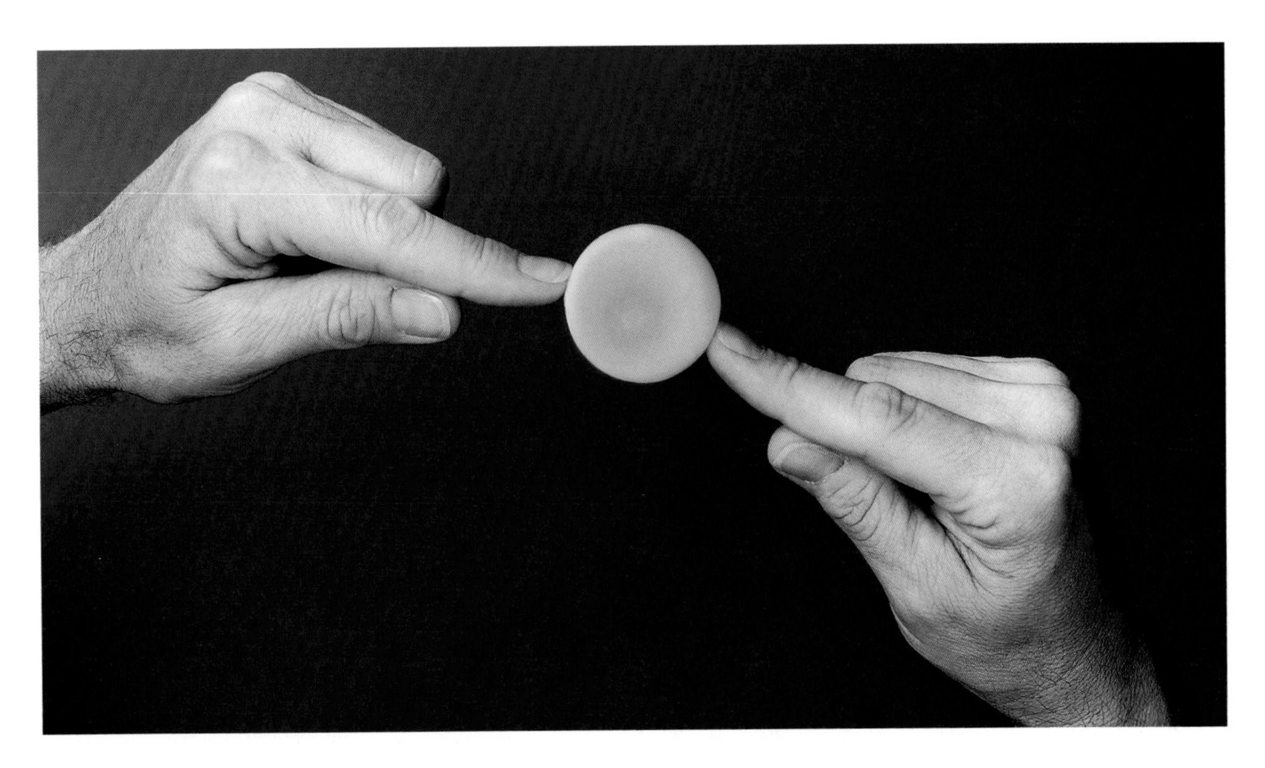

FIGURE 5.9: Energy Ball

HAVING A BALL

With the Energy Ball (Figure 5.9), you can explore closed and open circuits. In the experiment, you made a circle with your friends. That made the energy ball work because the circle of people made a way for the electrons to get from one part of the Energy Ball to the other. When someone broke the circle by letting go of hands, the electrons could not move to the other part of the Energy Ball. That's why the ball turned off. It works the same way with all electrical circuits. If the circuit is all connected, the electric charges can move. Then we say that the circuit is *closed*. If there is a break in the circuit, it is called an *open circuit* (Figure 5.10).

The Energy Ball is actually more complicated than that. The electrons that traveled through your friends are not the same ones that lit up the ball. When the Energy Ball detects that electrons are moving from one part of the ball to the other part, it turns on a second circuit to make the light and sound.

Closed circuit

Open circuit

FIGURE 5.10: Closed and open circuits.

Web Resources

Circuits for beginners.
www.bbc.co.uk/schools/scienceclips/ages/6_7/electricity.shtml

A simulation in which students explore why balloons stick to clothes. Images show charges in a sweater, balloons, and the wall.
http://phet.colorado.edu/en/simulation/balloons

Explore electric charges around the house with a simulation starring John Travoltage.
http://phet.colorado.edu/en/simulation/travoltage

Compare an electric current to the flow of water.
www.cabrillo.edu/~jmccullough/Applets/Flash/Electricity%20and%20Magnetism/Water-Analogy.swf

See how a light switch opens and closes a circuit.
www.cabrillo.edu/~jmccullough/Applets/Flash/Electricity%20and%20Magnetism/Light-Switch.swf

Try to make a circuit to light up a lightbulb.
http://phet.colorado.edu/en/simulation/circuit-construction-kit-dc-virtual-lab

See how the Energy Ball works.
https://sites.google.com/site/sed695b3/projects/discrepant-events/closed-circuit

Relevant Standards

Note: The Next Generation Science Standards *can be viewed online at* www.nextgenscience.org/next-generation-science-standards.

PERFORMANCE EXPECTATIONS

3-PS2-3

Ask questions to determine cause and effect relationships of electric or magnetic interactions between two objects not in contact with each other. [Clarification Statement: Examples of an electric force could include the force on hair from an electrically charged balloon and the electrical forces between a charged rod and pieces of paper....]

4-PS3-2

Make observations to provide evidence that energy can be transferred from place to place by sound, light, heat, and electric currents.

DISCIPLINARY CORE IDEAS

PS3.A: Definitions of Energy

- Motion energy is properly called kinetic energy; it is proportional to the mass of the moving object and grows with the square of its speed. (MS-PS3-1)
- A system of objects may also contain stored (potential) energy, depending on their relative positions. (MS-PS3-2)
- Temperature is a measure of the average kinetic energy of particles of matter. The relationship between the temperature and the total energy of a system depends on the types, states, and amounts of matter present. (MS-PS3-3), (MS-PS3-4)

PS3.B: Conservation of Energy and Energy Transfer

- When the motion energy of an object changes, there is inevitably some other change in energy at the same time. (MS-PS3-5)
- The amount of energy transfer needed to change the temperature of a matter sample by a given amount depends on the nature of the matter, the size of the sample, and the environment. (MS-PS3-4)
- Energy is spontaneously transferred out of hotter regions or objects and into colder ones. (MS-PS3-3)

6

ELECTRIC CIRCUITS

All electrical gadgets and gizmos need an electric current to work. To make a gadget work correctly, the electrical charges have to be steered to the right places. This is why electric circuits are needed. With a very simple circuit, you can light a lamp. More complicated circuits make a television work, or a remote control, or an automatic dishwasher.

Each gadget presented here is made of parts called *components*. Every component has a role in making the gadget work. One component might make the right amount of electricity. Another component might switch a light on and off or make a sound from a speaker. To make it easier for you to learn about and make circuits, you will need to know the symbols for some components. In this chapter, you will learn about these symbols and then plan and build simple electric circuits. You will also measure the electric current using basic units of electricity.

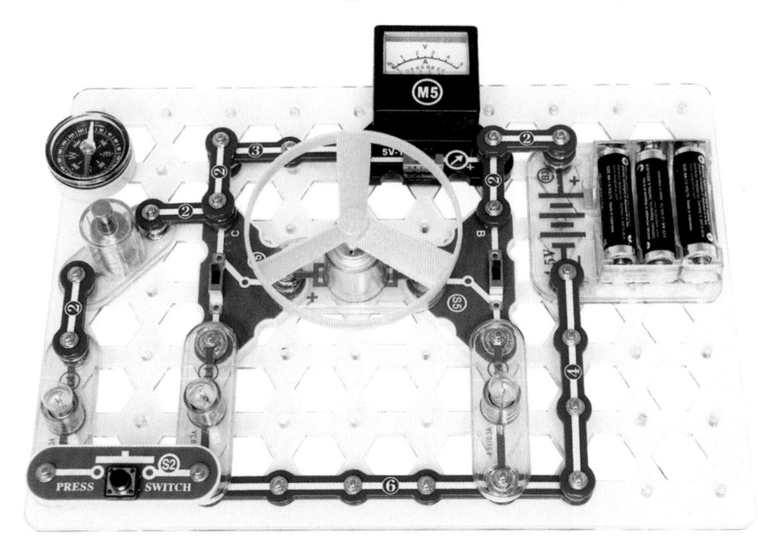

6 Let's Explore!

FIGURE 6.1: Snaptricity

SAFETY NOTES

- Read and follow the safety instructions in the Snaptricity instruction manual.
- One of the experiments in this chapter involves a food item. Never eat any food used in classroom experiments.

SNAPPING CIRCUITS

Snaptricity (Figure 6.1) is a box of electric components that are easy to use—just snap them together. In these experiments you will learn more about electric circuits, connections, and components.

Preparations for the Experiments

1. Carefully read the safety instructions in the Snaptricity instruction manual.
2. What gadgets do you have? Explore and identify the parts in the box.
3. Get to know the symbols shown in Figure 6.2. Explain the symbols with the help of the internet or the instruction manual included in the Snaptricity kit. The symbols are used to make it easier to model real circuits. They will also help you build circuits with others in your group.

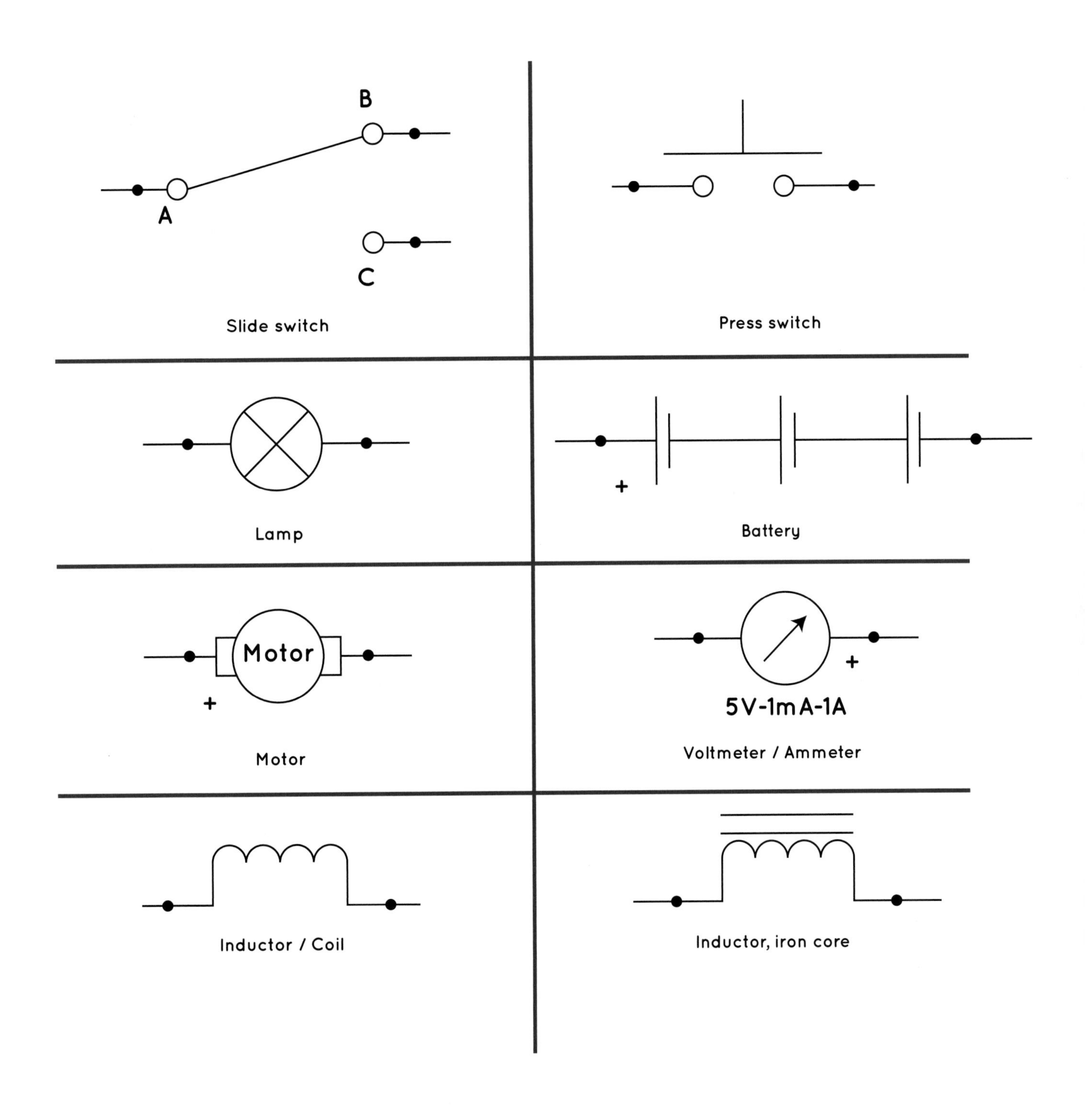

FIGURE 6.2: Symbols representing parts of a circuit.

LIGHTING A LAMP

1. Build a circuit so that a lamp lights up (Figure 6.3). You will need a battery, a lamp, and snap wires or jumper wires. (Loose wires used to connect components are sometimes called *jumper wires.*)

Hint: **Create a path for the current starting from one of the battery's poles to the other pole.**

2. Make a circuit in which you can switch the lamp on and off (Figure 6.4). You will need a battery, a lamp, a press switch, and snap wires or jumper wires. With the press switch you can turn on the lamp when you push the button. When you let go of the button, the lamp will turn off.

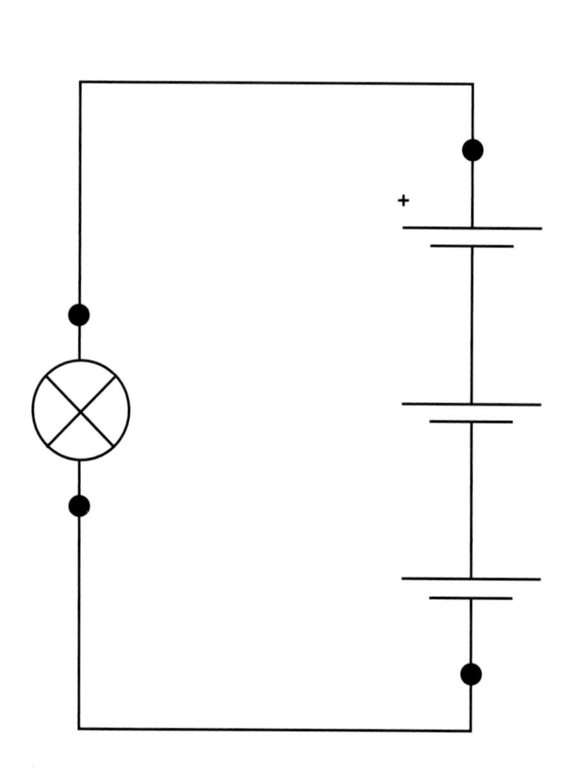

FIGURE 6.3: A circuit that allows you to light up a lamp.

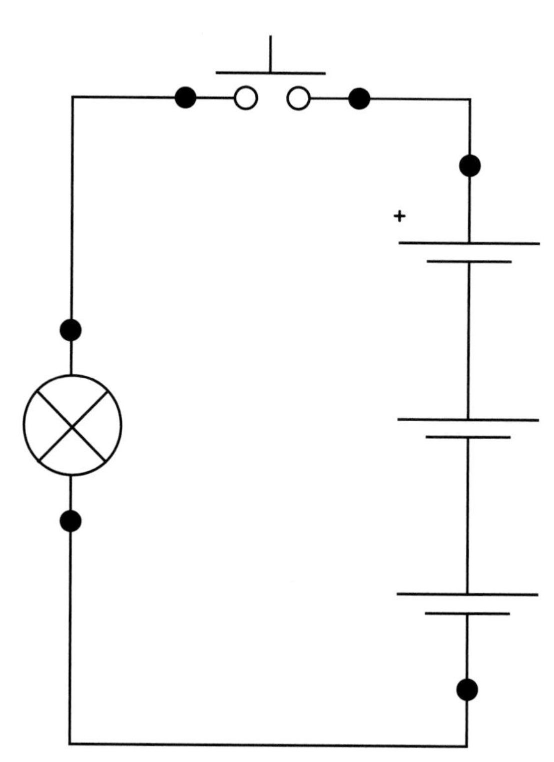

FIGURE 6.4: A circuit that allows you to turn a lamp on and off.

GETTING TO KNOW THE SWITCHES

1. Your task now is to control two lamps with a switch (Figure 6.5). The circuit has to work so that when one light is on the other is off. The slide switch will connect point A to point B or point C. Check the components needed from the electric diagram below.
2. Replace one lamp with a motor and add a press switch to the circuit (Figure 6.6). When you turn the slide switch, you select the component you want to turn on (the lamp or the motor), and with a press switch you turn the power on.

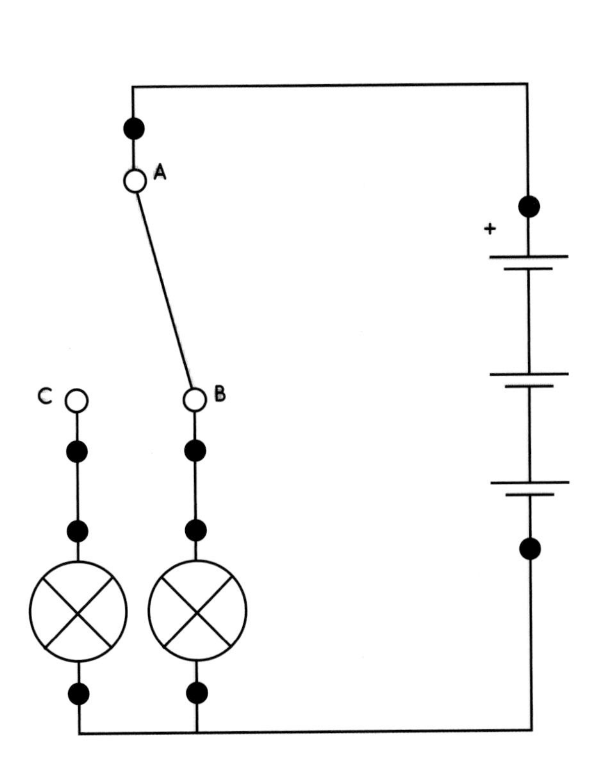

FIGURE 6.5: A circuit to control two lamps with one switch.

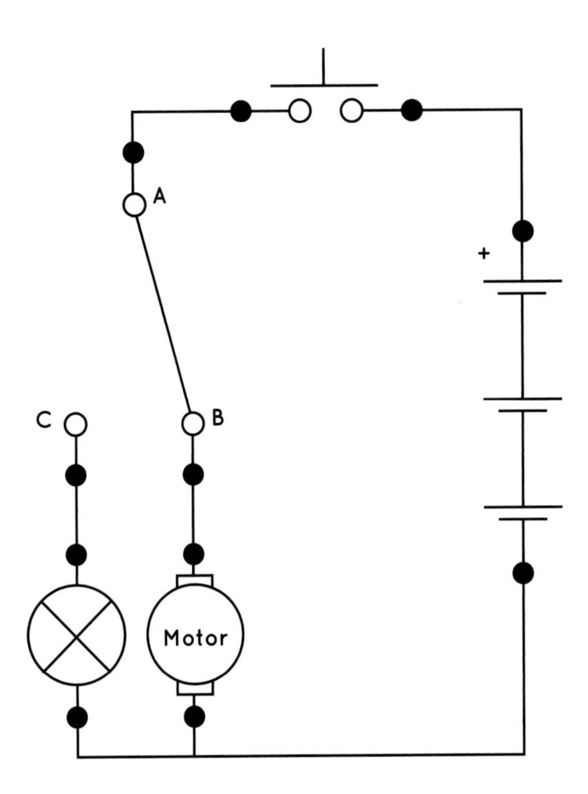

FIGURE 6.6: A circuit that allows you to choose which component to turn on.

LEARNING TYPES OF CONNECTIONS

1. Connect lamps in *series.* This means connecting them like railroad cars—one after another. Start with one lamp in a circuit (Figure 6.7).
2. Observe the brightness of the lamps when you add the second (Figure 6.8) and the third lamp (Figure 6.9).
3. What happens to the brightness when you add lamps?
4. What happens if you snap one of the lamps out of the circuit?
5. Next, you will connect lamps in *parallel.* In a parallel connection, the lamps are next to each other, like cars parked in a supermarket parking lot. Follow the electric diagram (Figure 6.10) and build the circuit.
6. Again observe the brightness as you add more lamps to the circuit.
7. What happens to the brightness when you add lamps?
8. What happens if you snap one of the lamps out of the circuit?

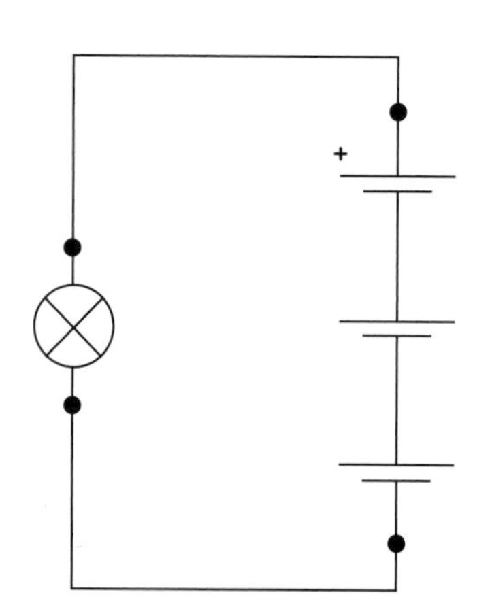

FIGURE 6.7: Start your lamp series with one lamp in the circuit.

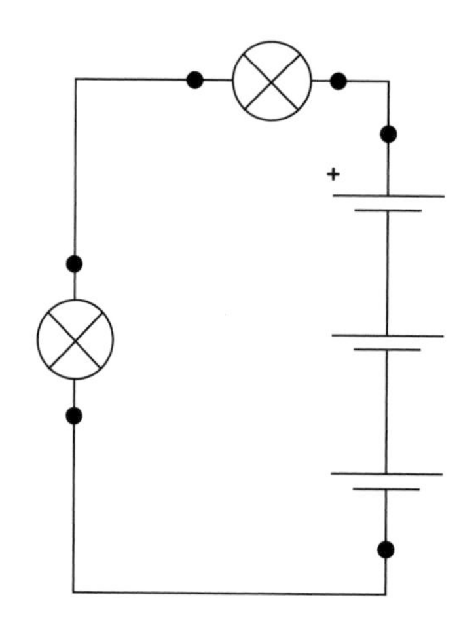

FIGURE 6.8: Add a second lamp to the circuit.

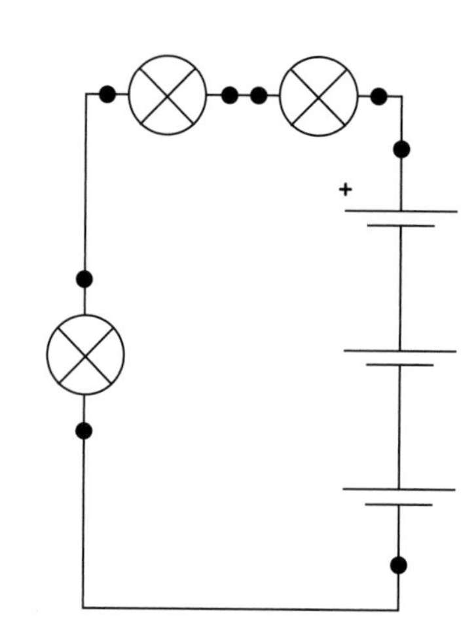

FIGURE 6.9: Add a third lamp to the circuit.

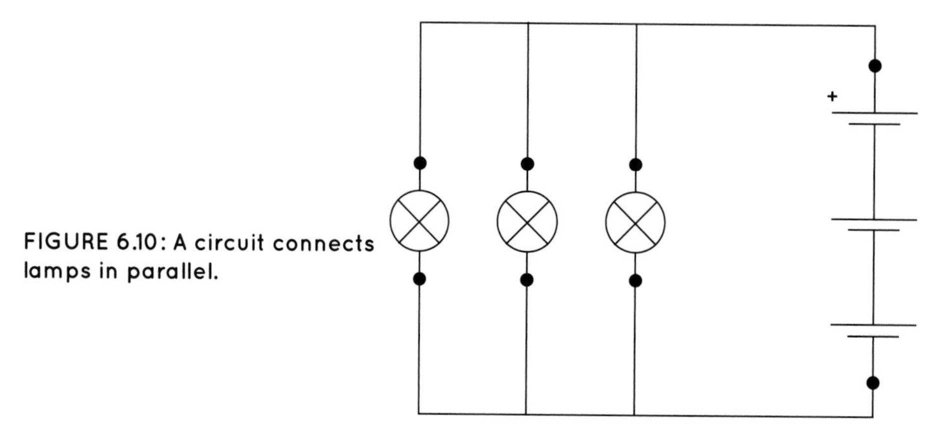

FIGURE 6.10: A circuit connects lamps in parallel.

EXPLORING CONDUCTIVITY

1. With the next circuit, you will explore electrical conductivity. Build a circuit like the one in Figure 6.11.
2. Put different items between the ends of the jumper wires. You can try materials such as metal, paper, wood, glass, and so on.
3. Make a list of the materials that conduct electricity and those that do not. (*Hint*: It might help to refresh your memory on open and closed circuits.)

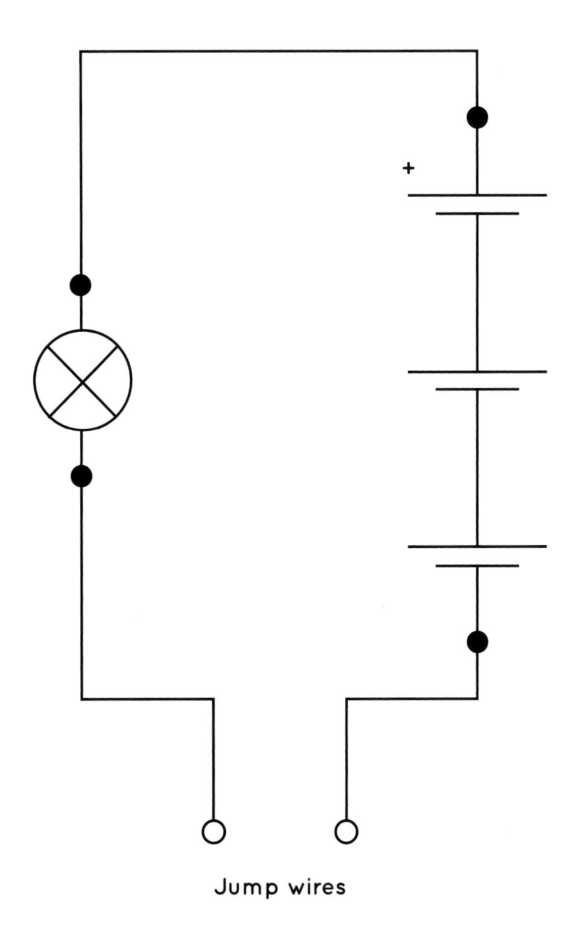

FIGURE 6.11: A circuit with jumper wires.

MEASURING ELECTRICITY

You may have heard the words *voltage* and *current*. The voltage in a battery creates an electric current when you have a closed circuit. A current is a flow of electric charges in a circuit. The greater the voltage in the battery, the greater the current it creates. The voltage is given in units called *volts* and can be measured with a *voltmeter*. Electrical current is measured in units called amperes. Electrical current is measured with an *ammeter*.

Volts

1. In the next experiment, you will measure the voltage in batteries using a voltmeter. Attach jump wires to the meter. The red wire should be connected to the positive (+) snap and the black wire to the other snap. Select 5V from the meter's slide switch so that you are measuring volts. Take one battery and touch its positive pole with the end of the red wire and touch its negative pole with the end of the black wire (Figure 6.12). Now read the battery's voltage on the meter. How much is it?
2. Next, add one battery so that the two batteries in the circuit are in series (like connected railroad cars; Figure 6.13). How many volts do two batteries create? How about three batteries in series (Figure 6.14)?

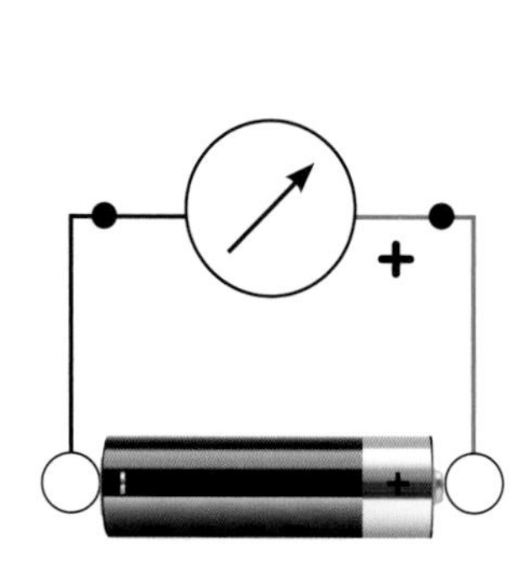

FIGURE 6.12: Touch the battery's positive pole with the red wire and the negative pole with the black wire.

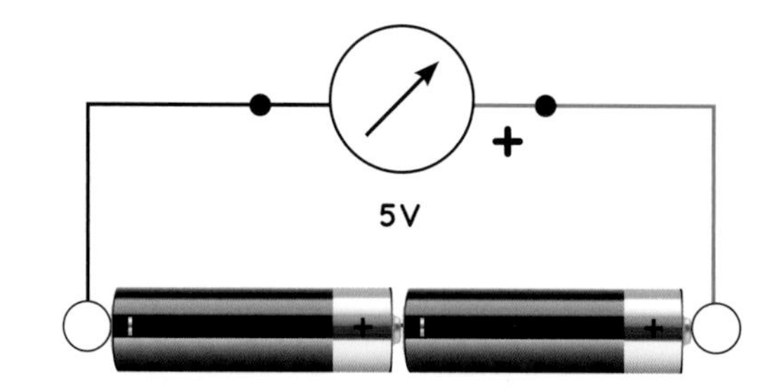

FIGURE 6.13: Connect a second battery in series.

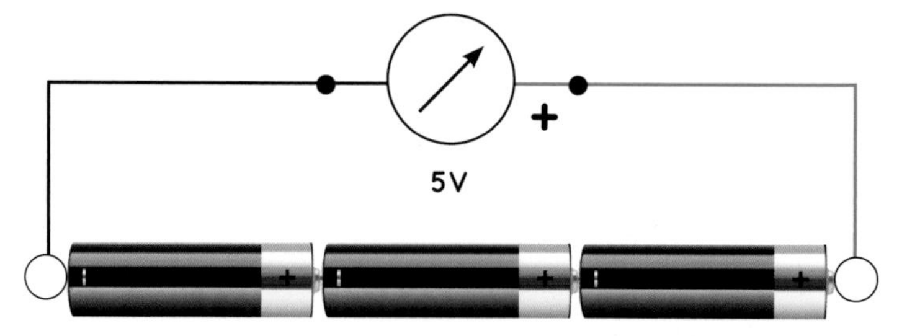

FIGURE 6.14: Connect a third battery in series.

Amperes

1. With the same meter, you can also measure small currents, such as one thousandth of an ampere (called one milliampere or 1 mA). You can also measure bigger currents—up to one ampere (1 A). For this next experiment, select the 1 A scale with the meter's slide switch.
2. Build a circuit in which you can turn the light on by pressing the switch. Then measure the current in the circuit with the meter (Figure 6.15). *Hint*: You will need to put the ammeter into the circuit so that the current must pass through it.

What is the reading of the meter?

3. Add another lamp in series (Figure 6.16). How did the reading change?

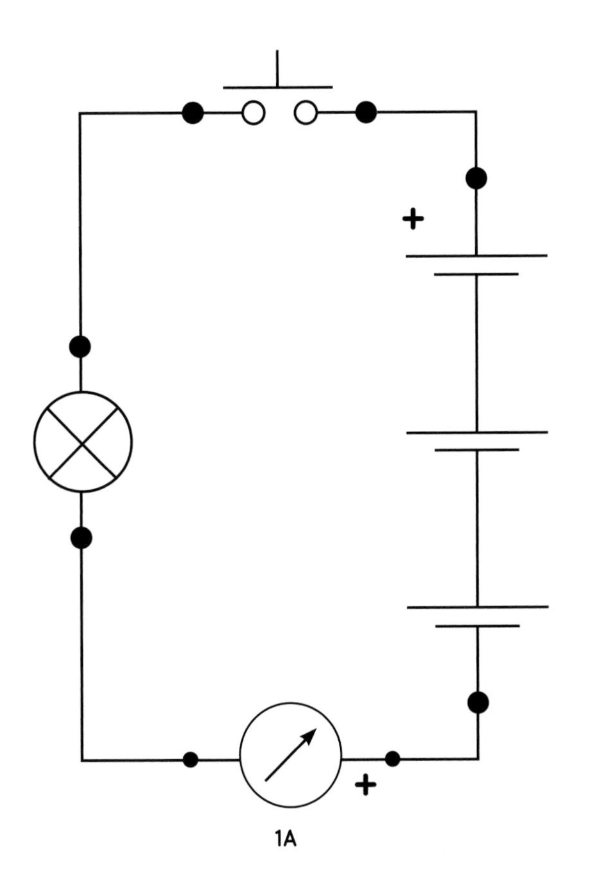

FIGURE 6.15: Measure the current of a one-lamp circuit with an ammeter.

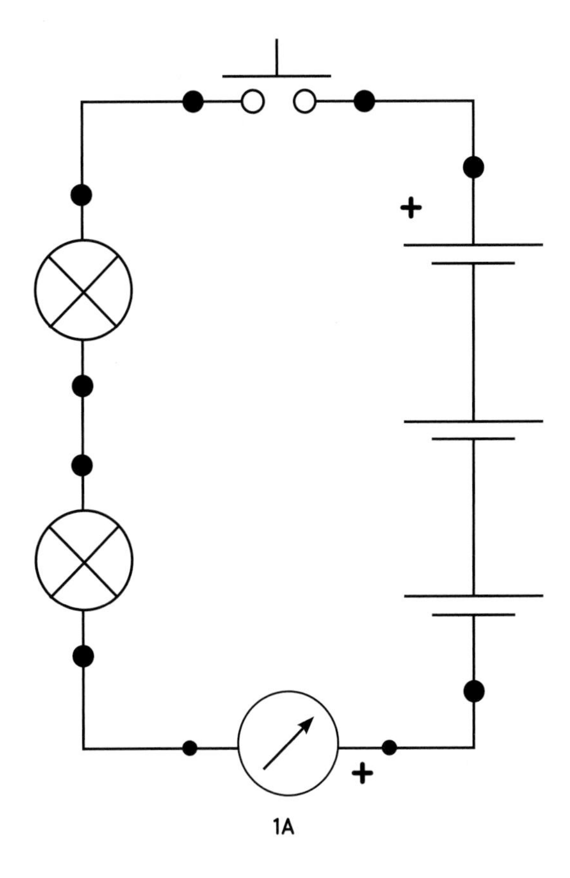

FIGURE 6.16: Measure the current of a two-lamp circuit with an ammeter.

CREATING VOLTAGE WITH A SIMPLE BATTERY

1. You learned that there is a voltage between the ends of a battery. In a battery the voltage is created by a chemical reaction. You can make a potato and some metal strips work like a battery. Here's how: Pick a moist vegetable or fruit. It could be an orange, lemon, avocado, cucumber, potato, or something else. In a conventional battery, there are two different metals and a liquid that is *conductive*. This means that the electrical current can flow through the liquid.
2. See if you can figure out how to create a voltage with a vegetable or fruit, two jumper wires, and two different metal strips. Measure the voltage with the meter. Keep in mind that there has to be a circuit for there to be an electrical current—so build a circuit!
3. (*Optional*) If you can find another pair of metal strips—from another Snaptricity box—see if you can increase the voltage by using a second vegetable or fruit.

CREATING VOLTAGE WITH A HAND CRANK

1. The Hand Crank (Figure 6.17) will create a voltage just like a battery. Use the Hand Crank to light up a lamp.
2. Use the Hand Crank to run the motor with the propeller attached to it.

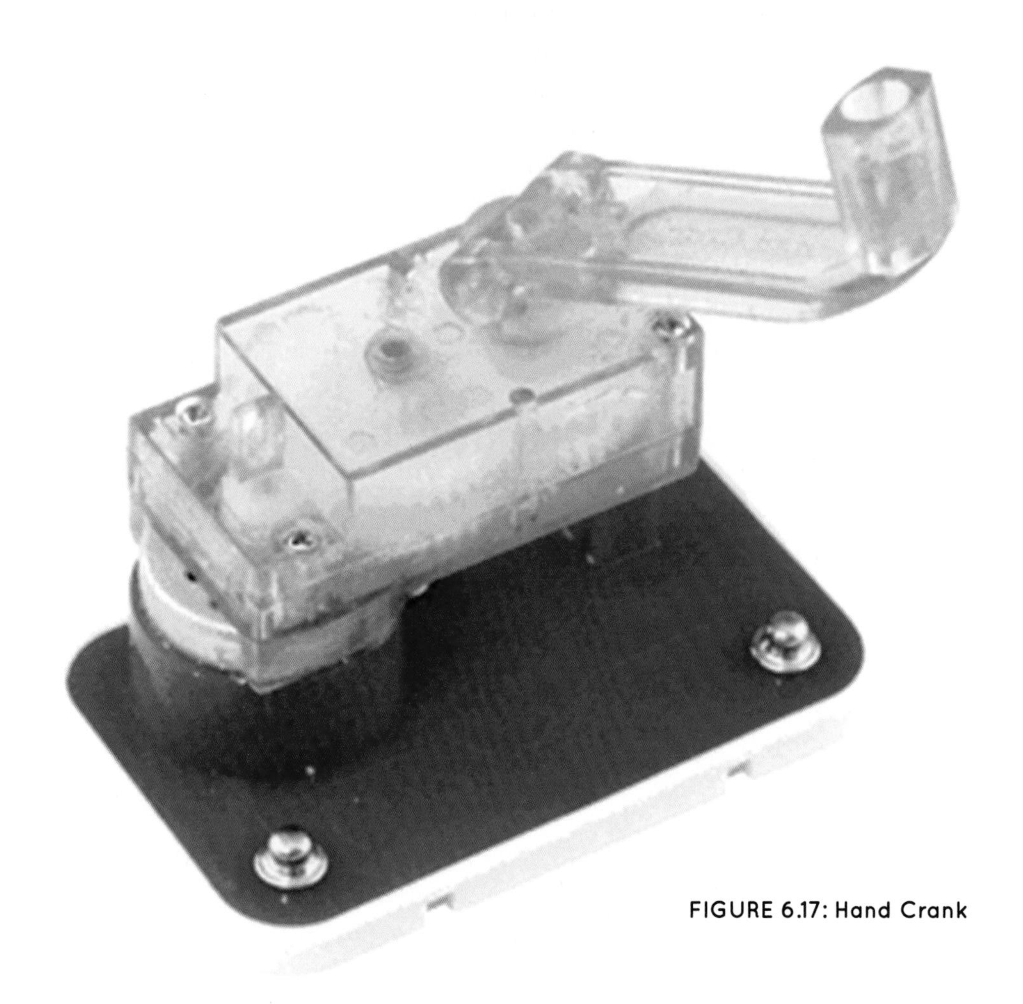

FIGURE 6.17: Hand Crank

6 What's Going On?

ELECTRIC CIRCUITS

With a battery you can create an electric current in a closed circuit.

An electric current will light up a lamp when the current travels through the lamp. When the circuit is open, there is no electric current flowing in the circuit. Your first experiment was to light up a lamp (Figure 6.18).

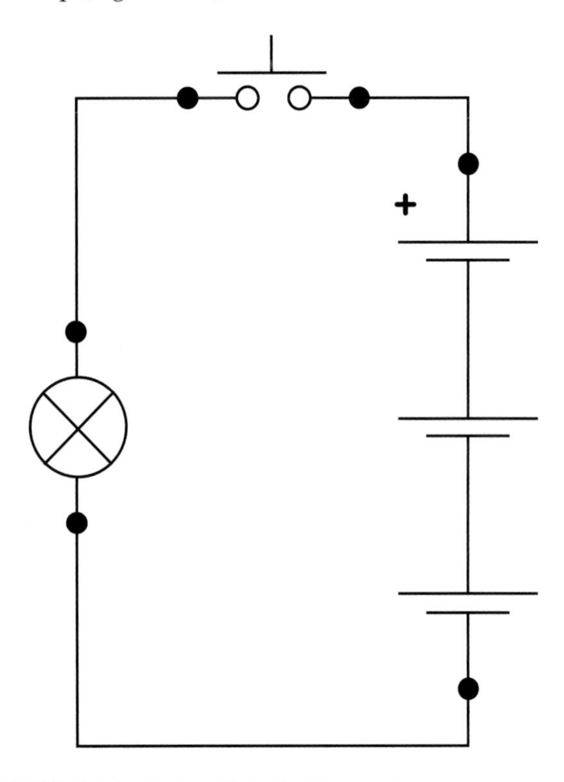

FIGURE 6.18: A circuit to light up a lamp.

With a switch you can close or open the circuit. When you press the switch, you connect the wires to close the circuit.

With the slide switch, you choose which way the electric current will flow: As shown in the Figure 6.19, it will either make the motor work or light up a lamp.

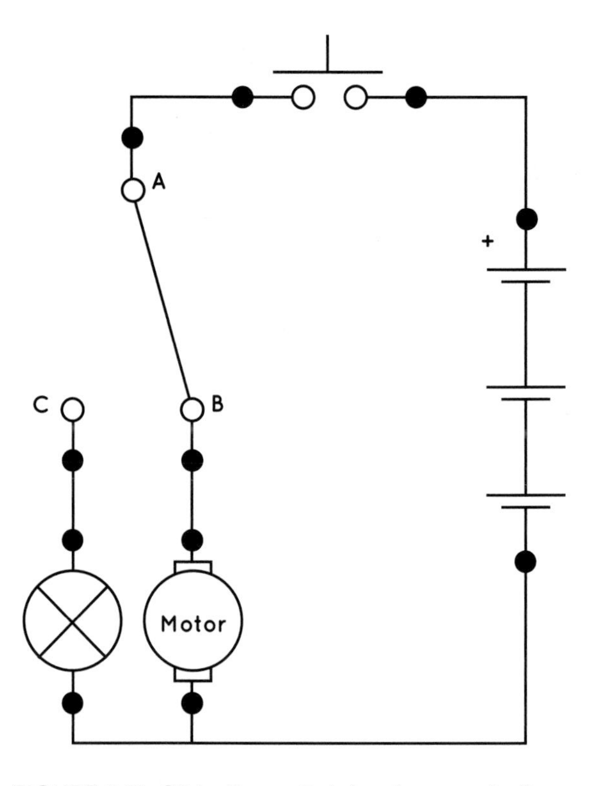

FIGURE 6.19: Slide the switch to choose whether the lamp or the motor will come on.

Types of Circuits

Series Circuit

When more lamps are connected in series (Figure 6.20), the brightness of the lamps will decrease. This is because adding more lamps to the circuit reduces the current and voltage through each individual lamp.

When snapping off one of the lamps, you should have noticed that all the lamps go off. You have created an open circuit.

Parallel Circuit

When the lamps are connected in parallel, their brightness does not change (Figure 6.21). That is because each lamp is connected directly to the battery, and each lamp gets the full voltage from the battery.

When snapping off one of the lamps, you can see that the other two lamps are still on.

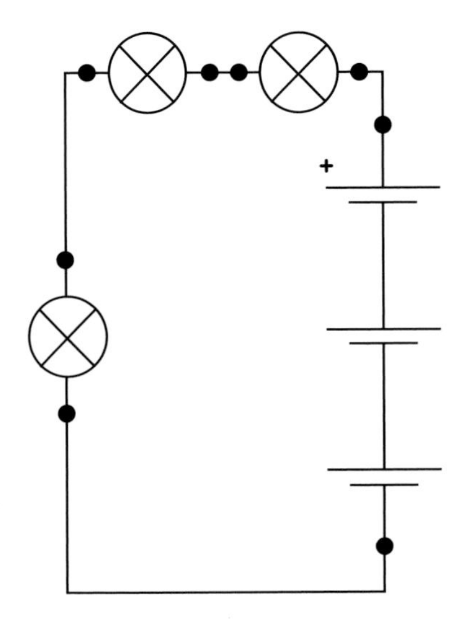

FIGURE 6.20: Lamps connected in series.

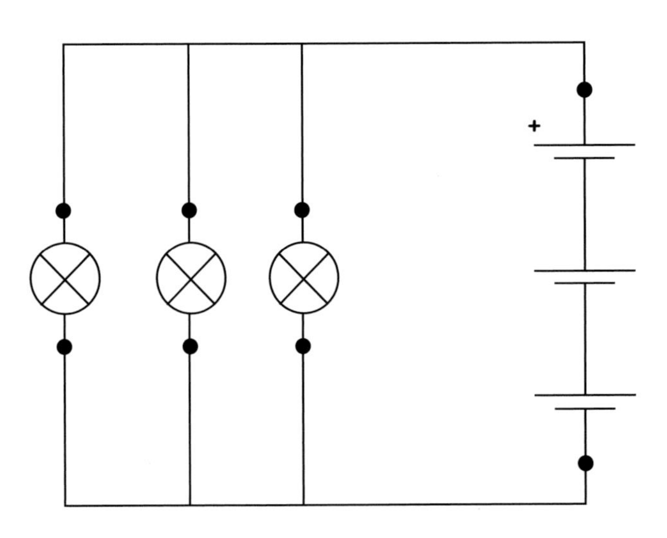

FIGURE 6.21: Lamps connected in parallel.

CONDUCTIVITY

Materials that conduct electricity are called conductors. If you had a conductor connected to the jump wires, then the light lit up. Most metals are good conductors. If a material does not conduct electricity, it is called an *insulator*. Wood, paper, and rubber are good insulators.

BASIC UNITS OF ELECTRICITY

The battery's voltage creates a current in a closed circuit. A current is a flow of electric charges in a circuit. In the circuits you make here, the greater the voltage of the battery, the greater the current will be.

One AA battery has voltage of 1.5 volts. When you connect batteries in series, you increase the voltage. The new voltage is the sum of the voltages from the individual batteries. The voltage for two batteries is 3 volts and for three it is 4.5 volts.

Current is a flow of electric charges in a circuit. It is measured in units of amperes. The greater the reading, the greater the current and the brighter the lamp. When you have only one lamp and the meter in the circuit, the reading is approximately 0.3 amperes. When you added a second lamp in the circuit, the current reading decreased.

If you are able to do the experiment with the fruit or vegetable, you can increase the voltage in your fruit or vegetable circuit by using jumper wires to connect a second fruit or vegetable in series.

SNAPTRICITY SETUPS FOR THE CIRCUITS IN THIS CHAPTER

Note that your circuits do not have to look exactly like these. They just have to work!

Lighting a Lamp

Light up one lamp.

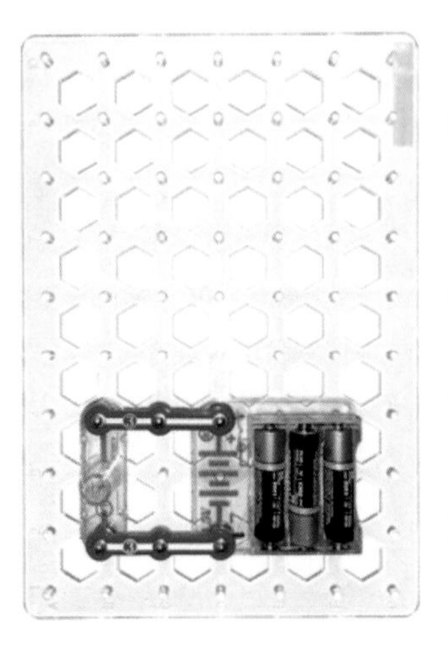

Light up the lamp using a circuit with a switch to turn it on.

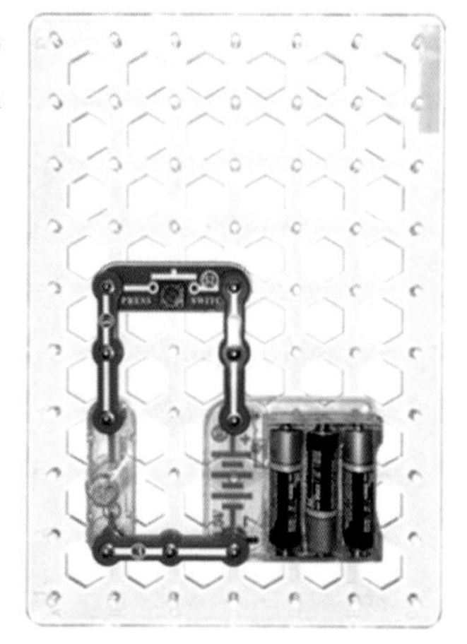

Or create a circuit using wires.

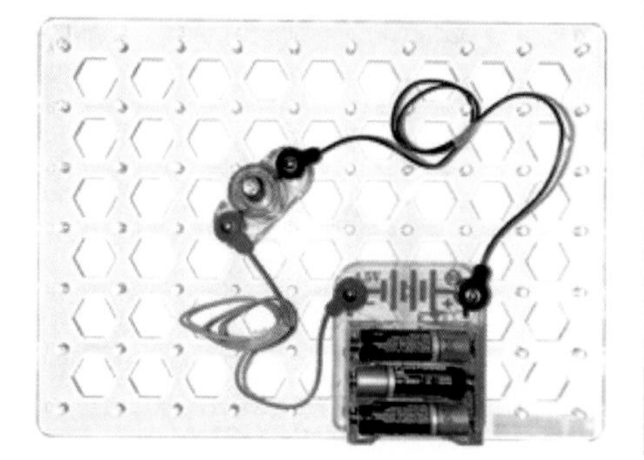

Getting to Know the Switches

Control two lamps with a switch.

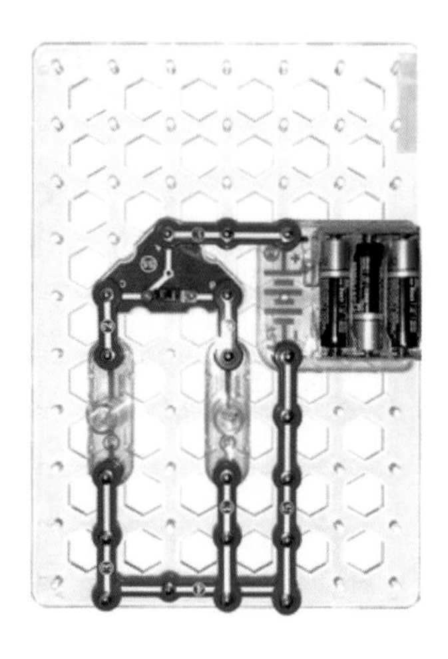

Replace one lamp with a motor, and add a press switch to the circuit.

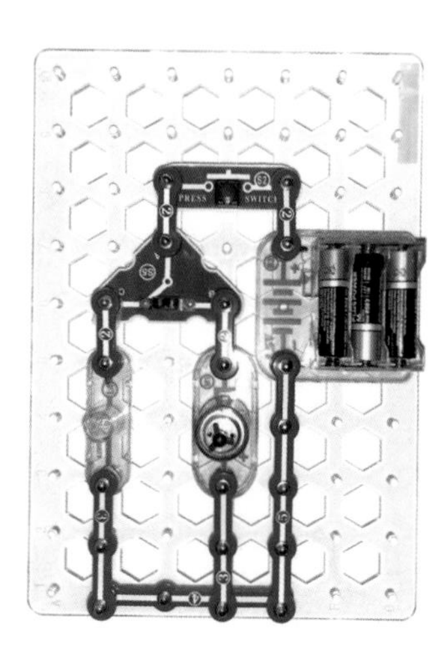

Learning Types of Connections

Connect lamps in *series*. Start with one lamp in a circuit.

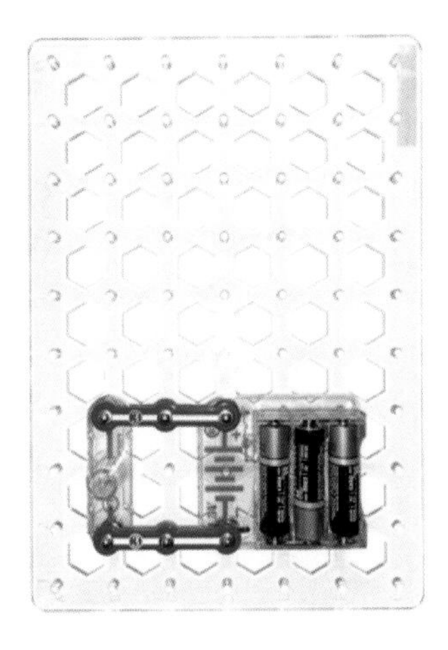

Observe the brightness of the lamps when you add the second and the third lamp.

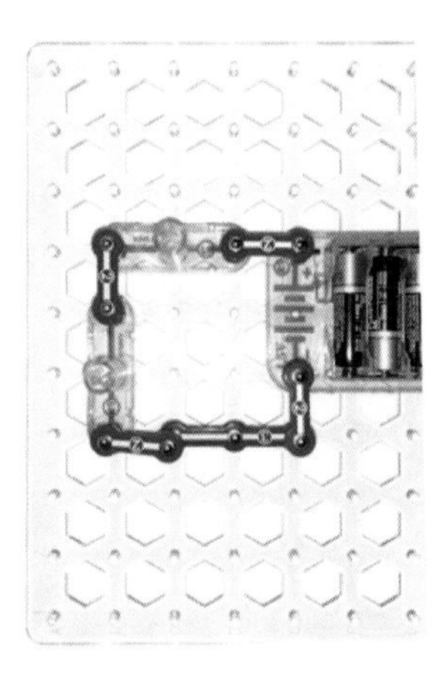

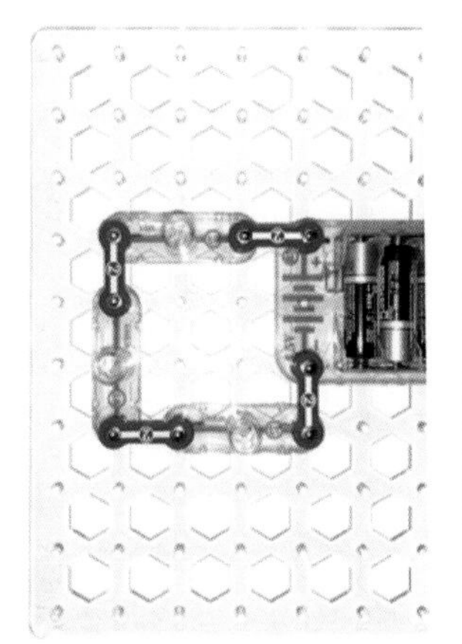

Connect lamps in *parallel*. One, two, and three lamps.

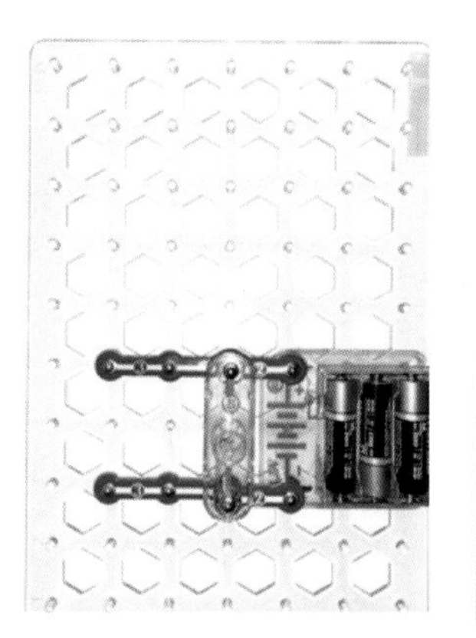

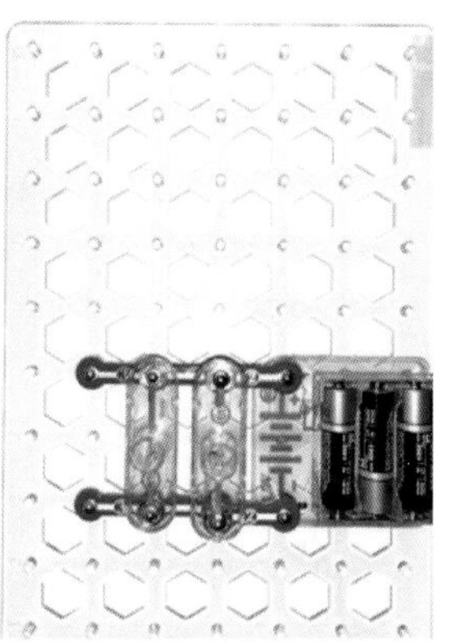

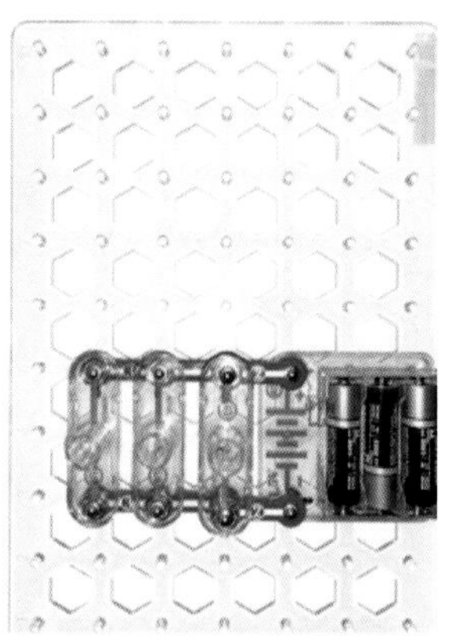

Exploring Conductivity

Use this circuit to explore electrical conductivity.

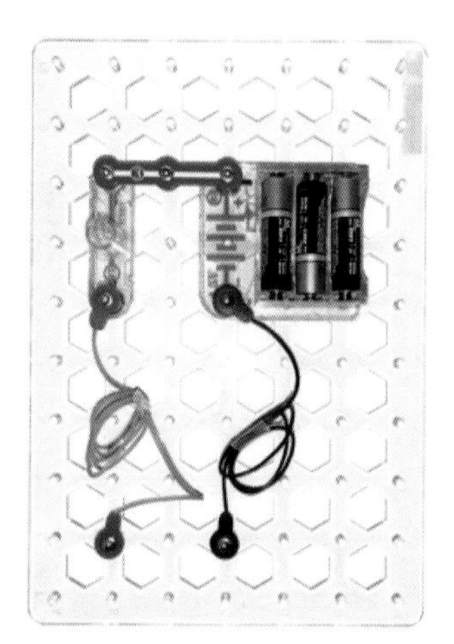

Measuring Electricity

Measure the voltage in batteries using a voltmeter. Attach jumper wires to the meter.

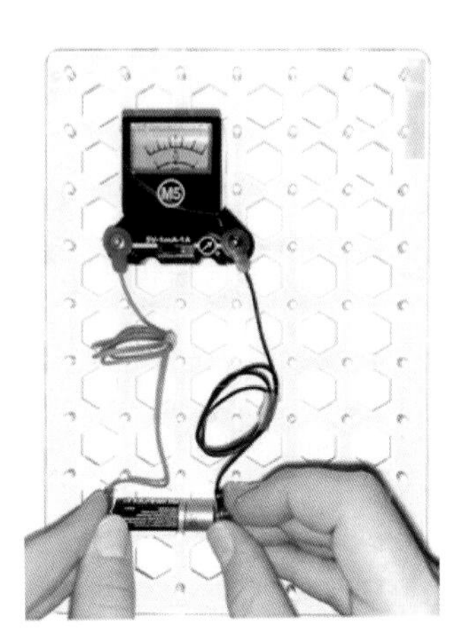

Next, add another battery so that the two batteries in the circuit are in series.

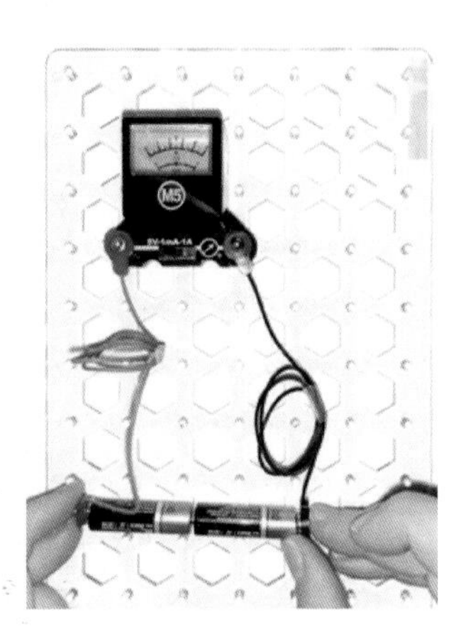

How about three batteries in series?

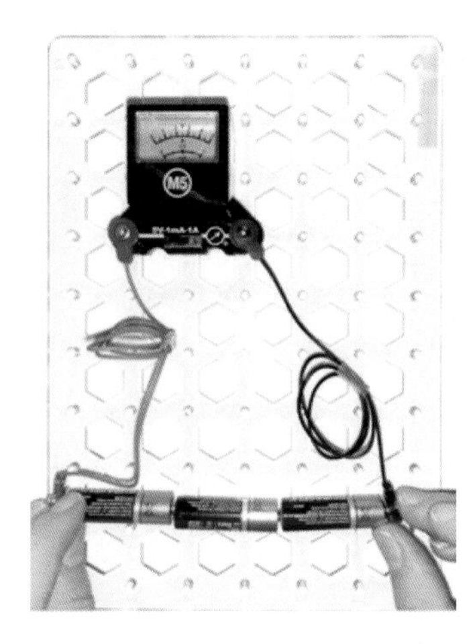

Build a circuit in which you can turn the light on by pressing the switch. Then, measure the current in the circuit with the meter.

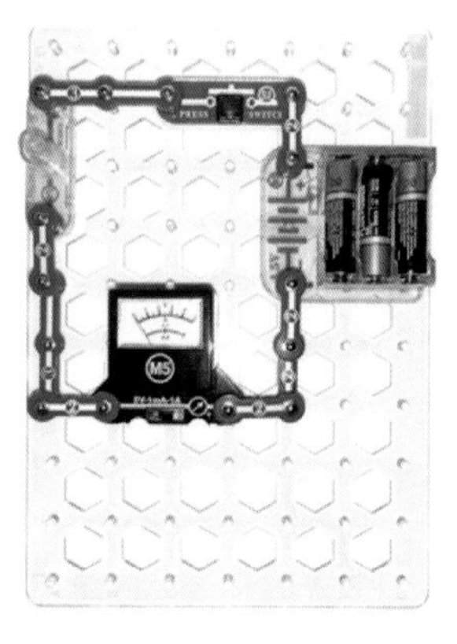

Add another lamp in series.

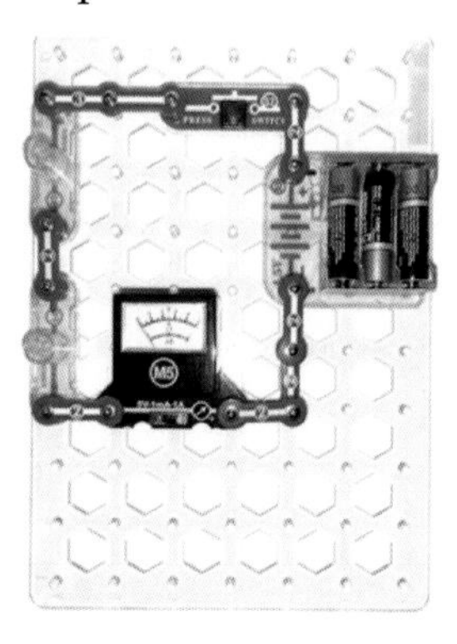

Creating Voltage With a Simple Battery

This battery is made from two metal strips and a potato!

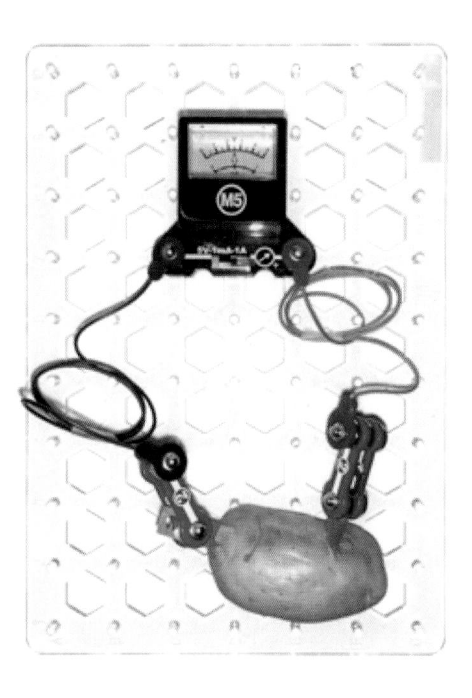

Creating Voltage With a Hand Crank

Use the Hand Crank to light up a lamp.

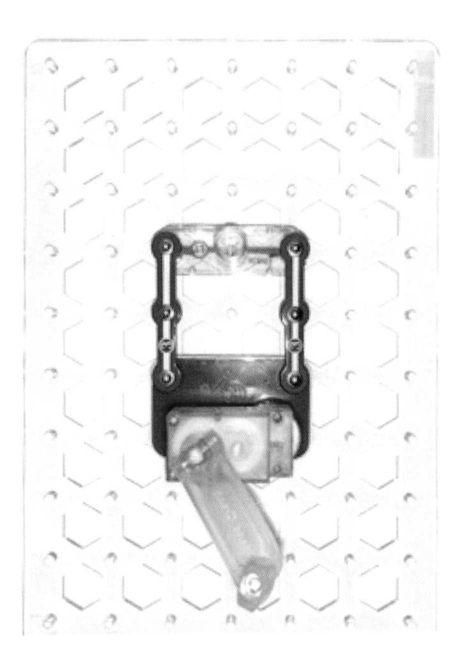

Use the Hand Crank to run the motor with the propeller attached to it.

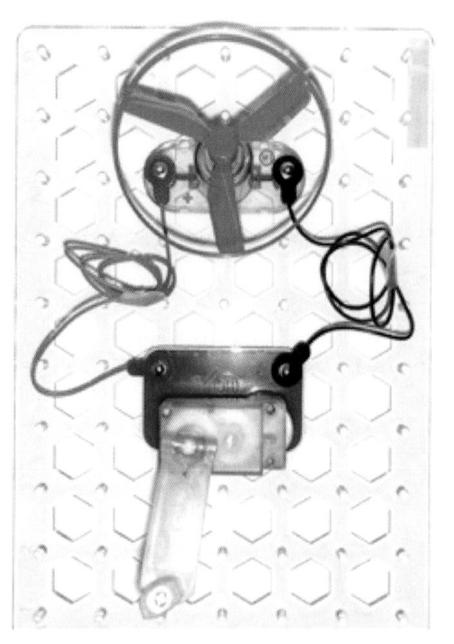

Web Resources

These PDF manuals contain many projects that can be done with Snap Circuits.

Projects 1–101: *www.elenco.com/admin_data/pdffiles/753102.pdf*

Projects 102–305: *www.elenco.com/admin_data/pdffiles/753098.pdf*

Projects 306–511: *www.elenco.com/admin_data/pdffiles/753104.pdf*

Projects 512–692: *www.elenco.com/admin_data/pdffiles/753292.pdf*

Manual for Snap Circuits Light, Projects 1–182.
www.elenco.com/admin_data/pdffiles/753285.pdf

Another Snaptricity manual with 78 projects.
www.elenco.com/admin_data/pdffiles/753303.pdf

Relevant Standards

Note: The Next Generation Science Standards *can be viewed online at* www.nextgenscience.org/next-generation-science-standards.

PERFORMANCE EXPECTATION

4-PS3-2

Make observations to provide evidence that energy can be transferred from place to place by sound, light, heat, and electric currents.

DISCIPLINARY CORE IDEAS

PS3.A: Definitions of Energy

- Energy can be moved from place to place by moving objects or through sound, light, or electric currents.

PS3.B: Conservation of Energy and Energy Transfer

- Energy is present whenever there are moving objects, sound, light, or heat. When objects collide, energy can be transferred from one object to another, thereby changing their motion. In such collisions, some energy is typically also transferred to the surrounding air; as a result, the air gets heated and sound is produced.
- Light also transfers energy from place to place.
- Energy can also be transferred from place to place by electric currents, which can then be used locally to produce motion, sound, heat, or light. The currents may have been produced to begin with by transforming the energy of motion into electrical energy.

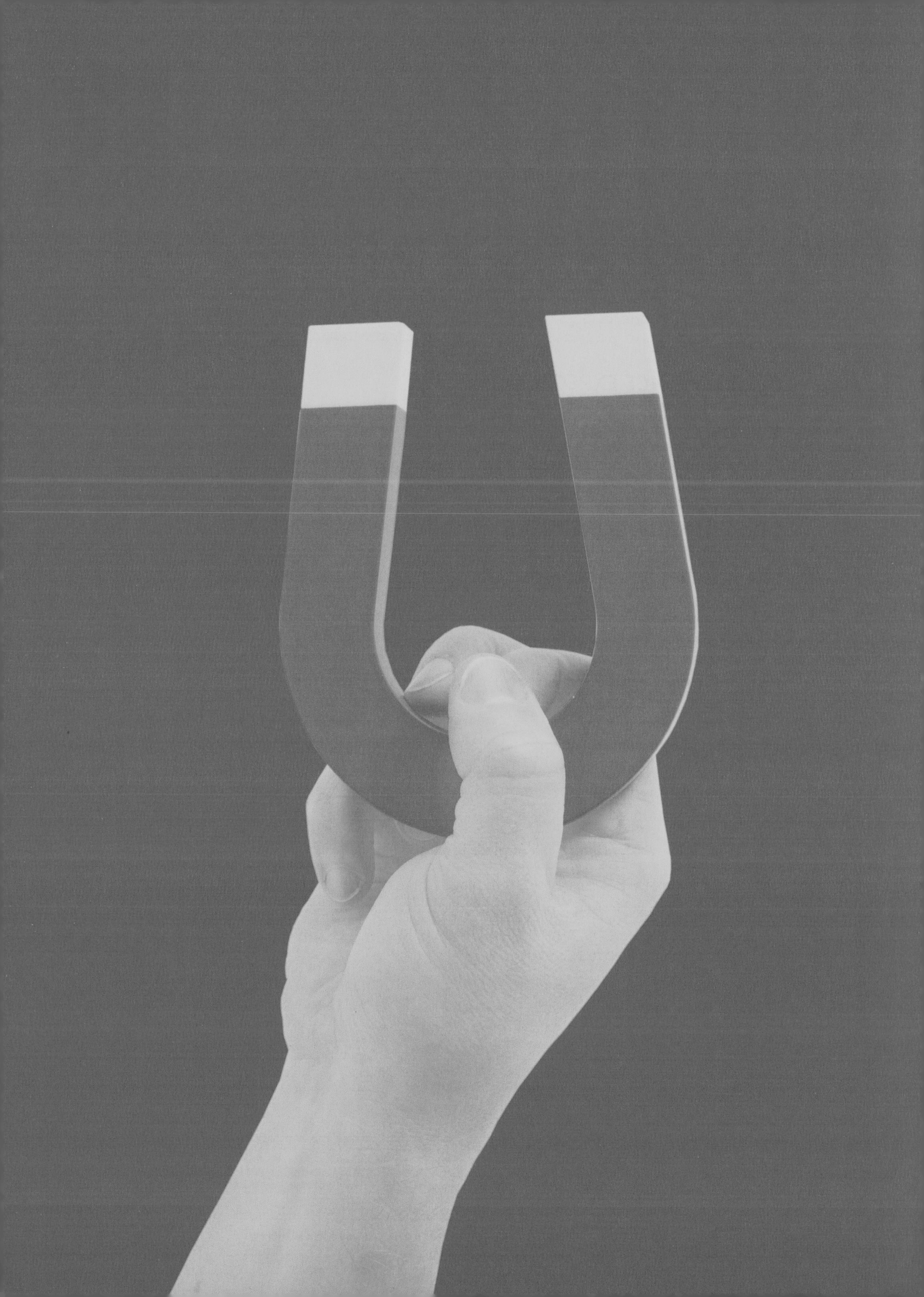

MAGNETISM

You probably know that a magnet attracts objects made of iron. The phenomenon called magnetism has many uses. Maybe there are some magnets holding notes on the door of your refrigerator. Every motor in your home, such as the ones in a fan or a washing machine, uses magnetism—even your toy car that runs on batteries and your doorbell. A magnetic compass uses magnetism. Compasses were widely used to show which way to go long before people had GPS navigation systems.

In this chapter, you will explore magnetism in different ways.

7 Let's Explore!

SAFETY NOTES

- Wear safety glasses or goggles.
- Use caution in handling glass, metal, and so on, which can be sharp and cut or puncture skin.

BAR MAGNETS

Magnetism is a very interesting phenomenon, and playing with magnets is a lot of fun. You can find a bar magnet (Figure 7.1) in every Snaptricity box.

1. There are two letters written on the bar magnet found in the box. What do the letters N and S stand for?
2. You will need two bar magnets in the next experiment. You can borrow one more from a neighboring group and then lend yours to them. (If there is only one bar magnet, you can use the magnet in the 3D Magnetic Compass.) The ends of magnets are called *magnetic poles*. What happens if you move an N pole close to an S pole? What if you put two N poles near each other? Describe your observations.
3. Magnets do not attract all materials. Explore which materials are affected by a magnet. Try metals, wood, glass, plastic, rubber, and other materials you can find in your classroom.

FIGURE 7.1: A bar magnet with a south pole and a north pole.

MAGNETIC FIELD

Here, you will explore magnetic fields. A magnetic field is an area in which magnets affect other magnets and magnetic materials.

1. There is a box of iron filings (Figure 7.2) in the Snaptricity box. Iron filings look like a metal powder. What does the magnet do to the iron filings in the box?

FIGURE 7.2: Box of iron filings.

2. Take the compass (Figure 7.3) from the Snaptricity box and find out what it does when a bar magnet is brought close to it. Explain how the compass works.

FIGURE 7.3: Compass

3. Now use the 3D Magnetic Compass (Figure 7.4) and explore the magnetic field of a bar magnet with it. Search for other magnets in the classroom and explore the shapes of their magnetic fields.

FIGURE 7.4: 3D Magnetic Compass

4. Discuss what the 3D Magnetic Compass does when it is near different parts of the bar magnet. Does it matter how far it is from the bar magnet?

SAFETY NOTES

- Wear safety glasses or goggles.
- Wash hands with soap and water if directly exposed to iron filings.

MAGNETIC GLOBE

FIGURE 7.5: Magnetic Globe

Earth has a magnetic field around it. This magnetic field makes the compass needle turn to point toward the Earth's poles. Today we have a different kind of gadget that helps people know which way to drive their cars—a global positioning system (GPS) device. Using a map and a compass is an old-fashioned way to figure out which way to go, but it still works. And it doesn't even need electricity!

1. Explore the magnetic field of the Magnetic Globe (Figure 7.5) with the Magnetic 3D Compass. Describe the shape of Earth's magnetic field.
2. Locate where you live on the globe and check which way the magnetic field points there.
3. Use the 3D Magnetic Compass to find the actual direction of the magnetic field at your location. You can try this in your classroom, but it might work better outside the school and away from electric wires. Is the direction of the magnetic field where you are similar to the direction you found from the Magnetic Globe? If not, how they are different?

ELECTROMAGNET

You can create a magnet by sending an electric current through a coil of wire. This is called an electromagnet (Figure 7.6). In a coil of wire, the wire is all rolled up. Now it's time to make a magnet with electricity.

1. Build the circuit shown in Figure 7.7. See p. 53 if you need a refresher on component symbols.
2. Place the compass close to the coil that has an iron wire core inside. Then press the switch. What happens to the compass needle?
3. With the help of the electromagnet you just created, try lifting the thin iron rod that comes in the Snaptricity box.
4. Time for some competition! Put some paper clips on the table and see how many your electromagnet can pick up. How many paper clips can you lift using the electromagnet? Can anyone beat your best try?

FIGURE 7.6: Electromagnet

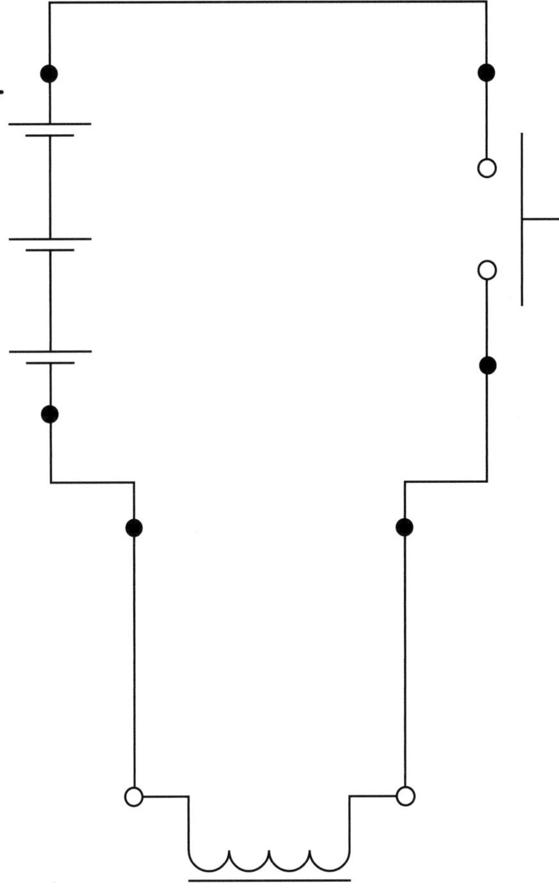

FIGURE 7.7: A circuit with an electromagnet.

7 What's Going On?

BAR MAGNETS

There are always two magnetic poles—N and S (Figure 7.8). N stands for the north pole and S for the south pole. If there are two magnets, the north poles push apart, or repel, each other. A north pole and a south pole attract each other.

With two bar magnets (Figure 7.9) you can see that two unlike poles (N and S or S and N) attract each other. Two like poles (N and N or S and S) repel each other, as shown in Figure 7.9.

Iron, steel, nickel, and cobalt are magnetic materials: They are attracted to magnets. They can also be magnetized. Coins are usually made of materials that are not magnetic.

FIGURE 7.8: Bar magnet

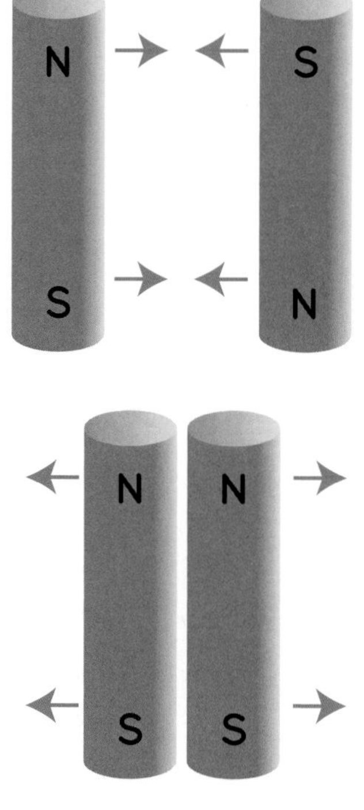

FIGURE 7.9: Opposite poles of a magnet attract each other. Like poles repel each other.

MAGNETIC FIELD

A magnetic field can go through nonmagnetic materials., so the iron filings in the box (Figure 7.10) are affected by the magnet. You can attract the filings through the plastic cover of the box. The filings tell you something about the magnetic field around the magnet. The filings move into the shape of the magnetic field.

FIGURE 7.10: Box of iron filings.

A compass (Figure 7.11) is one of the oldest uses of magnets. The needle in the compass is a little magnet. It has a north pole and a south pole as all magnets do. The north end of the compass needle is colored red. The compass needle turns easily, so the red, north end of the needle turns toward the bar magnet's south pole.

FIGURE 7.11: Compass

The Magnetic 3D Compass (Figure 7.12) has a gimbal-mounted bar magnet. "Gimbal-mounted" means that it can turn freely in any direction, so it can be used to find out the direction of a magnetic field, no matter what that direction is. It shows what direction the Earth's magnetic field is pointing. You can draw lines that show the direction of the magnetic field. These are called magnetic field lines. Figure 7.13 shows the magnetic field lines from a bar magnet. You can see if the little magnet in the 3D Magnetic Compass seems to point in the direction of the magnetic field lines drawn here.

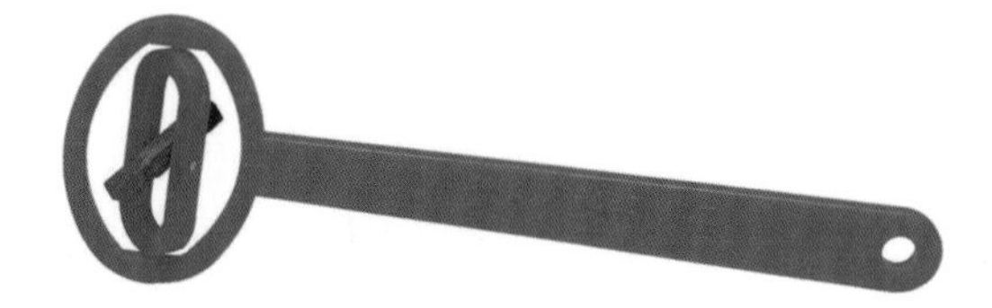

FIGURE 7.12: 3D Magnetic Compass

In the real world, there are not any magnetic field lines like this. These are just drawn here to show you the direction of the magnetic field.

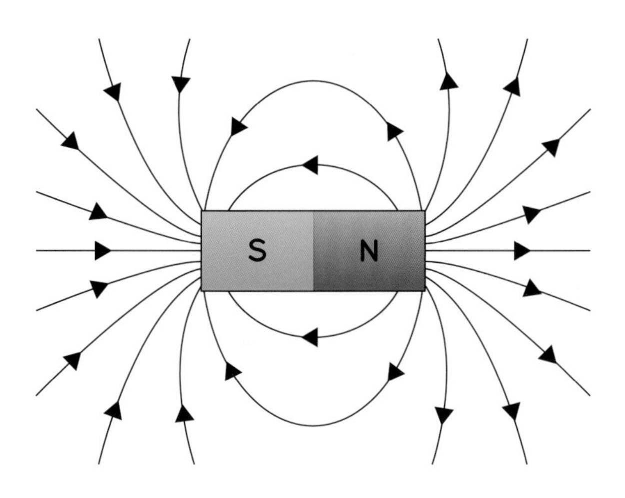

FIGURE 7.13: Magnetic field lines from a bar magnet.

FIGURE 7.14: Magnetic Globe

MAGNETIC GLOBE

In the Magnetic Field experiments you saw that the red north end of the needle in the compass turns toward the bar magnet's south pole. In the Magnetic Globe (Figure 7.14) experiments, you see that in Earth's magnetic field, the red end of the needle points north. That's because the Earth's south magnetic pole is located near Earth's geographic North Pole! You can see this in Figures 7.15 and 7.16 (p. 78).

The Earth does not really have a bar magnet inside, but the Earth's magnetic field is similar to the magnetic field from a bar magnet. To show this, we draw a bar magnet inside a picture of the Earth. Above is one way we might draw it.

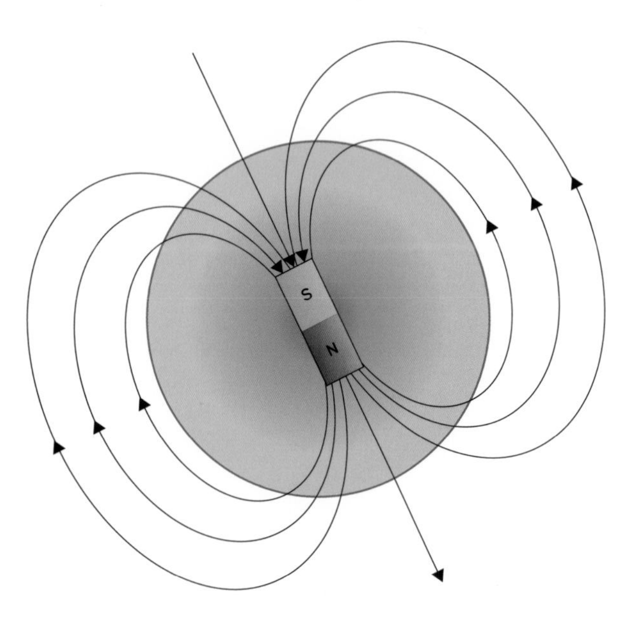

FIGURE 7.15: A bar magnet drawn inside a picture of the Earth shows that Earth's magnetic field is similar to that of a bar magnet.

Figure 7.16 shows another way to see how the Earth's magnetic field is like the field from a bar magnet. Notice how the bar magnet's south pole is near the Earth's North Pole. That is why the needle in a compass turns so that its red north end points toward the north. It is attracted to the Earth's south magnetic pole.

See how the compass in Figure 7.16 points toward the North Pole? Now you know why! It is pointing toward the Earth's south magnetic pole. With your 3D Magnetic Compass, you may notice that the compass points downward at some angle toward the ground. If you compare your position on the model you have—the magnetic globe—with your real place on Earth, you can understand why. The Earth is a giant globe, and if you could draw a straight line from your position to the North Pole, it would go through the ground.

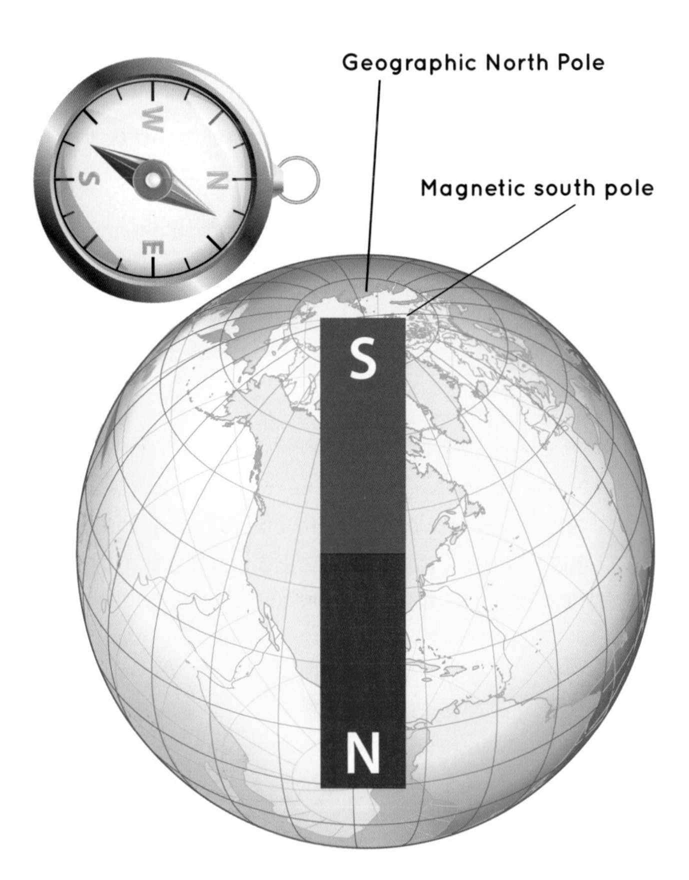

FIGURE 7.16: Another illustration showing how Earth's magnetic field is like the field from a bar magnet.

FIGURE 7.17: Electromagnet

ELECTROMAGNET

You can create a magnet with an electric current going through a coil of wire. An electromagnet (Figure 7.17) works the same way as a permanent magnet and attracts the same kinds of materials. When the circuit is closed, the electric current travels through the coil and makes the coil magnetic. The compass needle turns when you press the switch and close the circuit. You can even lift items like iron rods with it. Switching off the current makes the magnetic field disappear.

Your Snaptricity circuit with the electromagnet might look something like Figure 7.18.

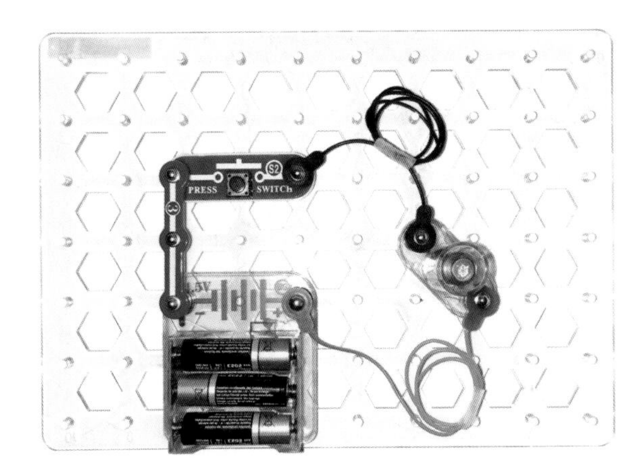

FIGURE 7.18: An electromagnet made with the Snaptricity kit.

Web Resources

A simulation allowing students to explore how a compass works, first with a bar magnet and then with the Earth's magnetic field.
http://phet.colorado.edu/en/simulation/magnet-and-compass

This simulation allows students to investigate how a compass and bar magnet interact, use a battery and some wire to make a magnet, play with the strength of the magnet, and more. (Don't forget to click on the Electromagnet tab.)
http://phet.colorado.edu/en/simulation/magnets-and-electromagnets

Magnet activities for elementary science.
www.internet4classrooms.com/science_elem_magnets.htm

Drag the compass needle to explore the magnetic field around the bar magnet. See the magnetic field line that goes through the compass needle.
www.windows2universe.org/physical_science/magnetism/bar_magnet_field_lines_java.html

Relevant Standards

Note: The Next Generation Science Standards *can be viewed online at* www.nextgenscience.org/next-generation-science-standards.

PERFORMANCE EXPECTATIONS

3-PS2-3

Ask questions to determine cause and effect relationships of electric or magnetic interactions between two objects not in contact with each other. [Clarification Statement: ... examples of a magnetic force could include the force between two permanent magnets, the force between an electromagnet and steel paperclips, and the force exerted by one magnet versus the force exerted by two magnets. Examples of cause and effect relationships could include how the distance between objects affects strength of the force and how the orientation of magnets affects the direction of the magnetic force.]

3-PS2-4

Define a simple design problem that can be solved by applying scientific ideas about magnets.* [Clarification Statement: Examples of problems could include constructing a latch to keep a door shut and creating a device to keep two moving objects from touching each other.]

5-PS1-3

Make observations and measurements to identify materials based on their properties. [Clarification Statement: … Examples of properties could include … response to magnetic forces.]

**The performance expectations marked with an asterisk integrate traditional science content with engineering through a Practice or Disciplinary Core Idea.*

DISCIPLINARY CORE IDEA

PS2.B: Types of Interactions

- Electric, and magnetic forces between a pair of objects do not require that the objects be in contact. The sizes of the forces in each situation depend on the properties of the objects and their distances apart and, for forces between two magnets, on their orientation relative to each other. (3-PS2-3), (3-PS2-4)

ENERGY

You need energy to stay alive. Energy comes to you in the form of food. Your body changes, or *transforms*, energy so that you can use it for activities such as walking, running, or riding a bike. The foods you eat—meat and vegetables—came from animals and plants, which also needed energy to grow. All that energy came from the Sun.

Energy is also needed to heat up a house. Heat is energy that is moving from a hotter place to a cooler place. For example, the energy moves out from a fireplace or a radiator.

Many electrical devices need batteries. Batteries are their source of energy. You can get energy from a wall outlet too. That energy is produced in an electrical power plant.

Energy is everywhere. It has several different forms. Energy can be moved from place to place in many ways. In the following examples you will find out more about what energy is, how it can be transported, and how is it possible to transform energy to different forms.

8 Let's Explore!

FIGURE 8.1: Super Solar Racer Car

SOLAR CAR

The energy from the Sun keeps living things alive. Plants use the energy from sunlight to help them grow. Sunlight can also be used to produce electricity. This is done with the help of solar panels. Here you will explore some of the forms of energy—and have a tight race with the Super Solar Racer Car (Figure 8.1).

1. Place your car in a sunny spot in the school yard.
2. Make a shadow to stop the car.
3. How does the energy move from the Sun to the Earth?
4. Make a track for cars where you can race against other groups. In the racetrack there should be several curves in one lap. You are allowed to steer only when the car is stopped! Measure the times.

Another idea: You can dim the lights in the classroom and try making the car work with a flashlight.

FIGURE 8.2: Ice Melting Blocks

MELTING ICE

Energy comes from the Sun in the form of light. That light hits the Earth and heats it. With the Ice Melting Blocks (Figure 8.2) you will explore another way to move, or *transfer*, energy.

1. Touch the surfaces of the two blocks. Which one feels warmer?
2. Make a prediction: On which block will an ice cube melt faster?
3. Put the O-rings on the blocks and set the ice cubes on the blocks too. Observe the melting of the ice for a few minutes. What do you see?
4. Touch the blocks again and sense the temperatures. Discuss what you found out.
5. Discuss possible reasons for why one ice cube melts faster.

***Tip:* If there are not enough setups for each group to try it, you can do this as a demonstration with a document camera.**

SAFETY NOTE

Immediately wipe up any splashed water to prevent a slip or fall hazard.

SAFETY NOTES

- Make sure there is nothing breakable in the vicinity.
- Wear safety glasses or goggles

FIGURE 8.3: Music Box

MUSIC BOX

Next you will explore moving energy with sound using the Music Box (Figure 8.3). This is a very important phenomenon because it makes hearing possible.

1. Figure out how the gadget works.
2. Wind up the gadget and then listen. What do you notice?
3. Place the music box on different surfaces, such as a blackboard, table, or window. How is the sound different now?
4. While the music box is playing on a table, put your ear against the table. What do you notice?
5. Try building a simple earphone: While the music box is playing, press it against your elbow and put your index finger in your ear.
6. Put the music box aside for a while and take a flyer (for example the butterfly) from the Fun Fly Stick box. Charge the flyer with the Fun Fly Stick and attach the flyer to a loudspeaker.
7. Play some music through the speaker, and turn up the volume—not too loud, you don't want it to hurt your ears!
8. Discuss in your group what you observe.
9. What do you think sound really is?

FIGURE 8.4: Hand Crank

HAND CRANK

With this Hand Crank (Figure 8.4), the energy of movement is changed to energy in the form of electricity. The crank creates a voltage and an electric current. This can be used to make an electrical gadget work. This is similar to the way that energy is created in power plants.

1. Create an electric circuit in which you have a hand crank and a lamp in series. Turn the crank.
2. What happens if you turn the crank faster?
3. What happens if you turn the crank in the opposite direction?
4. Replace the lamp with an electric motor and propeller. Can you make the propeller fly with the energy you create?

▼

SAFETY NOTE

Make sure any fragile equipment/materials are removed from the area when launching the propeller.

8 What's Going On?

Energy is everywhere. You know energy is present when you see moving objects or light, feel heat, or hear a sound. You need energy to make something move or to create light or sound. Scientists say that energy is the ability to do work. In the experiments, you saw how energy is transformed and how it can be moved from one place to another.

SOLAR CAR

You can see the light from the Sun and feel the heat that it creates. Energy from the Sun *radiates* to the Earth. That means the energy comes to Earth in the form of light. This way of transporting energy is called *radiation*. You can feel radiation from the Sun when you stand outside or lie on a beach on a sunny day.

The light, or radiation, from the Sun makes your skin warmer. Light energy has been changed into *thermal energy*. Thermal energy is the energy in an object that makes it feel warm or hot. At night the Earth is turned so that the other side of the Earth is facing the Sun. At that time we are in shadow. That is why nights are usually cooler than days—and of course darker, too.

FIGURE 8.5: Super Solar Racer Car

With the Super Solar Racer Car (Figure 8.5), you explored how radiation transfers energy and makes the car move. The energy of light is first changed to electrical energy in the *solar cells*, which are on top of the car. That electrical energy is then changed to energy of movement with the help of an electric motor.

You can have a class discuss how a solar-powered car is different from a gasoline-powered car. What happens to the energy after gasoline is put into the car? Energy in the gasoline goes into the car's engine. There the gasoline is burned. The energy is then changed to energy of movement as the car moves along the road.

Just as you have to add gasoline to a car, your body needs energy too. Where does your body get its energy? Your body gets its energy from the food you eat!

MELTING ICE

Radiation is not the only way to transport energy. With the Ice Melting Blocks (Figure 8.6) you got to know another way—conduction. Conduction of heat means that thermal energy is moving through some material. Some materials conduct heat better than others. Most metals are good conductors. The materials that don't conduct thermal energy very well are called *insulators*. A thick coat can be a good insulator. It keeps you warm because thermal energy from your body does not go through the coat very fast.

Many plastics, such as Styrofoam, are known to be good insulators. That's why hot drinks are often served in Styrofoam cups (Figure 8.7). The Styrofoam keeps the thermal energy from the drink from escaping through the cup. That's good for keeping your hands from getting too hot. And since the energy stays in the drink, it stays warm longer.

In the Melting Ice experiment, the metal block is a good conductor. When you touched it, it felt cooler because energy from your skin was quickly conducted away through the metal. But actually both blocks started at the same temperature—room temperature. Not much conduction occurred with the block that was an insulator. Since your finger couldn't lose much energy through the insulator, it didn't feel as cold on that block.

FIGURE 8.7: Hot drinks are often served in Styrofoam cups.

The ice cube on the metal block melted faster than the one placed on the insulator block. The heat needed to make the ice cube melt was conducted faster in the metal block.

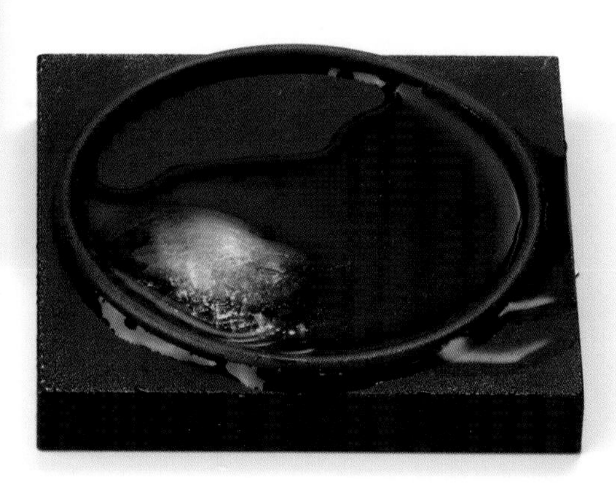

FIGURE 8.6: Ice Melting Blocks

MUSIC BOX

Sound is one form of energy. As it goes from one place to another, sound needs some material to travel through. It can move through air as well as in solid materials such as wood.

FIGURE 8.8: Music Box

When playing the Music Box (Figure 8.8), you made little metal strips *vibrate*, or shake back and forth very fast. You changed energy of movement into sound energy. The vibrating strips made the air vibrate. The sound you hear is actually a vibration of the air. The sound travels through the air to your ear. In your ear, the vibrating air makes your *eardrum* vibrate. In this picture of an ear, you can see where the eardrum is located (Figure 8.9).

The vibration of the air can be seen when a Fun Fly Stick flyer from Chapter 5 is attached to a speaker. It shakes, or vibrates, in time to the music.

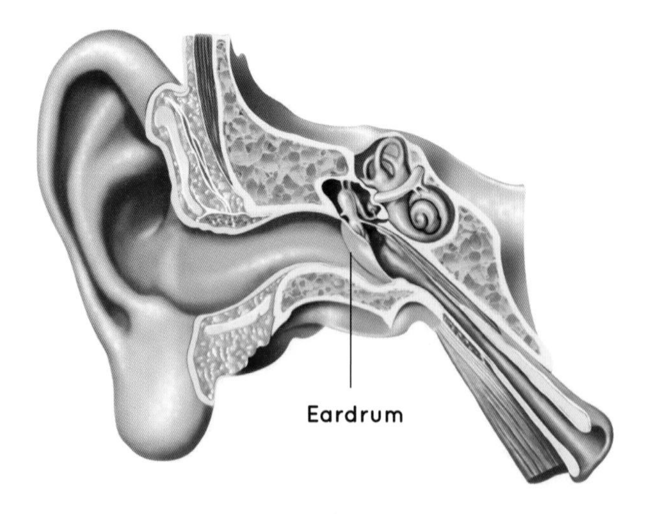

FIGURE 8.9: The location of the eardrum in an ear.

HAND CRANK

You used the Hand Crank (Figure 8.10) to transform the energy of movement into electrical energy. You learned earlier how electrical energy has many uses. Here you used it to create light energy (Figure 8.11) and to run a motor (Figure 8.12).

FIGURE 8.10: Hand Crank

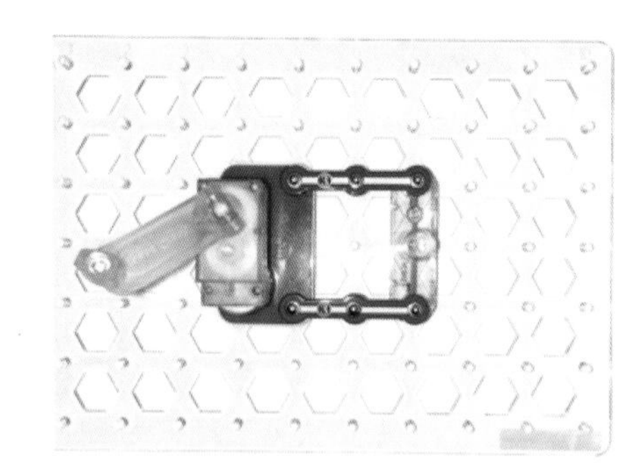

FIGURE 8.11: Hand Crank with light.

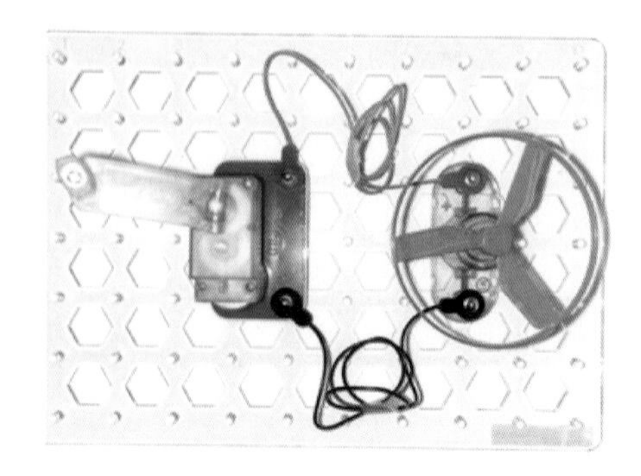

FIGURE 8.12: Hand Crank with motor and propeller.

Web Resources

A simulation to explore how heating and cooling iron, brick, and water adds or removes energy, and how energy is transferred between objects. Students can build a system that includes energy sources, changers, and users and track and energy as it moves through the system.
http://phet.colorado.edu/en/simulation/energy-forms-and-changes

Watch sound waves come out of a speaker.
http://phet.colorado.edu/en/simulation/sound

A simulation in which energy is generated using a bar magnet.
http://phet.colorado.edu/en/simulation/generator

Videos, lessons, and games about heat and temperature.
www.neok12.com/Heat-Temperature.htm

Solar cell simulation classroom activity.
www1.eere.energy.gov/education/pdfs/solar_cellsimulation.pdf

Relevant Standards

Note: The Next Generation Science Standards *can be viewed online at* www.nextgenscience.org/next-generation-science-standards.

PERFORMANCE EXPECTATIONS

K-PS3-1

Make observations to determine the effect of sunlight on Earth's surface. [Clarification Statement: Examples of Earth's surface could include sand, soil, rocks, and water] [Assessment Boundary: Assessment of temperature is limited to relative measures such as warmer/cooler.]

K-PS3-2

Use tools and materials to design and build a structure that will reduce the warming effect of sunlight on an area.* [Clarification Statement: Examples of structures could include umbrellas, canopies, and tents that minimize the warming effect of the sun.]

1-PS4-1

Plan and conduct investigations to provide evidence that vibrating materials can make sound and that sound can make materials vibrate. [Clarification Statement: Examples of vibrating materials that make sound could include tuning forks and plucking a stretched string. Examples of how sound can make matter vibrate could include holding a piece of paper near a speaker making sound and holding an object near a vibrating tuning fork.]

4-PS3-2

Make observations to provide evidence that energy can be transferred from place to place by sound, light, heat, and electric currents. [*Assessment Boundary: Assessment does not include quantitative measurements of energy.*]

4-PS3-4

Apply scientific ideas to design, test, and refine a device that converts energy from one form to another.* [Clarification Statement: Examples of devices could include electric circuits that convert electrical energy into motion energy of a vehicle, light, or sound; and, a passive solar heater that converts light into heat…]

**The performance expectations marked with an asterisk integrate traditional science content with engineering through a Practice or Disciplinary Core Idea.*

DISCIPLINARY CORE IDEAS

PS4.A: Wave Properties

- A simple wave has a repeating pattern with a specific wavelength, frequency, and amplitude. (MS-PS4-1)
- A sound wave needs a medium through which it is transmitted. (MS-PS4-2)

PS3.B: Conservation of Energy and Energy Transfer

- Energy is spontaneously transferred out of hotter regions or objects and into colder ones.

PS3.A: Definitions of Energy

- Temperature is a measure of the average kinetic energy of particles of matter. The relationship between the temperature and the total energy of a system depends on the types, states, and amounts of matter present.

CROSSCUTTING CONCEPT

Energy and Matter

- The transfer of energy can be tracked as energy flows through a designed or natural system.

APPENDIX

HOW TO ORDER THE GADGETS AND GIZMOS

Materials to support the lessons and experiments found throughout this book are available from Arbor Scientific in two NSTA Elementary School Physics Kits. Each kit includes 10–15 tools (listed below) to support lessons in velocity, friction, magnetism, pressure, energy, and more.

Visit *www.arborsci.com* to order the kits or learn more about their contents.

KIT #PK-0200	KIT #PK-0210
SPEED	**ELECTRICITY**
4 - Constant Velocity Car 1 - Velocity Radar Gun 4 - Pull-Back Car 4 - Stopwatch/Timer 4 - Windup Tape, 10 meters	4 - Fun Fly Stick 1 - Plasma Globe 4 - Energy Ball
FRICTION AND AIR RESISTANCE	**ELECTRIC CIRCUITS**
4 - Four-Sided Friction Block 4 - Spring Scale, 250g/2.5N Blue 4 - Air Puck 4 - Running Parachute	4 - Snaptricity kit
GRAVITY	**MAGNETISM**
4 - Balancing Bird 4 - IR-Controlled UFO flyer 1 - Vertical Acceleration Demonstrator	4 - 3D Magnetic Compass 4 - Magnetic Globe
AIR PRESSURE	**ENERGY**
4 - Atmospheric Pressure Cups, set of 2 4 - Vacuum Pumper and Chamber 1 - Air-Powered Projectile	4 - Ice Melting Blocks 4 - Super Solar Racer Car 4 - Hand Crank 4 - Music Box

CREDITS

IMAGES

MEDIAKETTU JARI PEURAJÄRVI

ARBOR SCIENTIFIC

RPM SPORTS

PAUL HEWITT

MATTI KORHONEN

MICHAEL GREGORY

SCOTT KANE

(p. 6: Radar gun, Scott Kane, http://scottkanephotography.com)

NIGEL TOUT

(p. 71: Birmingham International Maglev. Nigel Tout, Wikimedia Commons, CC BY-SA 3.0. *http://en.wikipedia.org/wiki/Maglev#mediaviewer/File:Birmingham_International_Maglev.jpg*)

OTHER CONTRIBUTIONS

OLIVIA BOBROWSKY

MARI MUINONEN AND HER STUDENTS

HANNU KORHONEN AND HIS STUDENTS

INDEX

*Page numbers in **boldface** type refer to tables or figures.*